NATURE LAW AND POLICY IN EUROPE

This volume considers current and future challenges for nature law and policy in Europe.

Following the Fitness Check evaluation of the Birds and Habitats Directives, in 2017 the EU adopted an Action Plan for nature, people and the economy to rapidly improve the Directives' implementation and accelerate progress towards the EU's biodiversity targets for 2020. More recently, the EU has adopted a Biodiversity Strategy for 2030 and proposed an EU Nature Restoration Law. This book makes a timely contribution by examining the current state of play in light of recent and historical developments, as well as the post-2020 nature law and policy landscape. While evidence suggests that Natura 2000 and the Habitats and Birds Directives have delivered conservation benefits for wildlife in Europe, biodiversity loss continues apace. The book reviews the requirements for an effective international nature conservation system, with reference to the Birds and Habitats Directives. It examines regulatory regimes, current legal issues in the fields of site protection and species protection, the protection of areas outside Natura 2000, recent developments in the EU and the UK, including the implications of Brexit, agriculture and nature conservation, litigation, science and access to justice. Written by leading experts in the field, from a range of stakeholder groups, the volume draws on diverse experiences as well as providing interdisciplinary perspectives.

This volume will be essential reading for students, scholars, practitioners, NGOs and policy-makers interested in European environmental policy and law, including for example lawyers, ecologists, environmental scientists, political scientists, natural resource managers, planners and civil servants.

Andrew L. R. Jackson is an Assistant Professor in Environmental and Planning Law in the Sutherland School of Law at University College Dublin, Ireland. He previously worked for the international law firm Slaughter and May in London and Paris; for the UK government's legal service, in the Department for Environment, Food and Rural Affairs (Defra); and with NGOs Friends of the Irish Environment and An Taisce, the National Trust for Ireland, where he was Natural Environment Officer and in-house Solicitor. He is the author of *Conserving Europe's Wildlife: Law and Policy of the Natura 2000 Network of Protected Areas* (Routledge, 2018).

Routledge Research in International Environmental Law

International Environmental Law and the International Court of Justice
Aleksandra Cavoski

Enforcement of International Environmental Law
Challenges and Responses at the International Level
Martin Hedemann-Robinson

Marine Pollution, Shipping Waste and International Law
Gabriela Argüello

Compensation for Environmental Damage Under International Law
Jason Rudall

International Environmental Law Compliance in Context
Mechanisms and Case Studies
Belen Olmos Giupponi

Sustainable Fisheries Management and International Law
Marine Fisheries in Bangladesh and the Bay of Bengal
Abdullah Al Arif

Marine Conservation and International Law
Legal Instruments for Biodiversity Beyond National Jurisdiction
Sarah Louise Lothian

Nature Law and Policy in Europe
Edited by Andrew L.R. Jackson

https://www.routledge.com/Routledge-Research-in-International-Environmental-Law/book-series/INTENVLAW

NATURE LAW AND POLICY IN EUROPE

Edited by
Andrew L. R. Jackson

Routledge
Taylor & Francis Group
LONDON AND NEW YORK

earthscan
from Routledge

Designed cover image: Micheál Ó Briain

First published 2023
by Routledge
4 Park Square, Milton Park, Abingdon, Oxon OX14 4RN

and by Routledge
605 Third Avenue, New York, NY 10158

Routledge is an imprint of the Taylor & Francis Group, an informa business

British Library Cataloguing-in-Publication Data
A catalogue record for this book is available from the British Library

ISBN: 978-0-367-27544-0 (hbk)
ISBN: 978-0-367-27997-4 (pbk)
ISBN: 978-0-429-29910-0 (ebk)

DOI: 10.4324/9780429299100

Typeset in Bembo
by codeMantra

Dedicated to the memory of Claus Stuffmann
and John Temple Lang

CONTENTS

ILLUSTRATIONS

FIGURE

TABLES

CONTRIBUTORS

Liam Cashman worked as an official in DG Environment of the European Commission between 1988 and 2021. In the first half of his career, he was the legal case handler for Ireland, dealing with environmental complaints and infringements, and contributing to relevant Commission court proceedings. In the second half, he worked on general policy initiatives related to access to justice and environmental compliance assurance, including environmental crime and environmental liability.

Guillaume Chapron is a Researcher at the Swedish University of Agricultural Sciences. Initially trained as a veterinarian, he has a doctorate in Ecology and a graduate diploma in Political Science. He uses a diversity of methods and disciplines, ranging from Bayesian state space models to legal analyses to answer questions at the core of contemporary conservation challenges, with particular emphasis on the conservation of large carnivores.

Carol Day holds an MSc in Nature Conservation and has worked in the environmental sector for 35 years, working for The Wildlife Trusts in Warwickshire and Surrey before moving to WWF UK. It was during this time that she qualified as a solicitor, spending over 20 years advising on countryside policy and legal issues. Since 2013 she has split her time between public interest law firm Leigh Day and the Royal Society for the Protection of Birds (RSPB). The Environment Team at Leigh Day specialises in a wide range of environmental cases including planning, wildlife law and climate change, including advising the recently formed NGO Wild Justice. Carol is also passionate about access to environmental justice and in her capacity with the RSPB, she co-chairs the Legal Strategy Group of Wildlife and Countryside Link and coordinates the Joint (UK) Link's work on access to justice. She is also a Trustee and Vice-Chair of the Environmental Law Foundation (ELF). She has lived in London for over 20 years, but still really misses the countryside.

Yaffa Epstein is an Associate Professor of Environmental Law at Uppsala University and a Pro Futura Scientia fellow at the Swedish Collegium for Advanced Study. Her current research foci are nature protection, the relationship between law and natural science, and the interaction between international, EU and national environmental law.

Ian Hodge is Professor Emeritus of Rural Economy in the Department of Land Economy at the University of Cambridge and a Life Fellow of Hughes Hall, Cambridge. He was Head of the Department of Land Economy between 2002 and 2011. Professor Hodge has a background in agricultural economics and rural planning and previously worked at the Universities of Newcastle upon Tyne and Queensland before moving to Cambridge in 1983. Professor Hodge is past President of the Agricultural Economics Society. His research and teaching is primarily in the areas of rural governance, land and environmental management, and agricultural and environmental policy.

Andrew L. R. Jackson is an Assistant Professor of Environmental and Planning Law at University College Dublin. He previously worked for Slaughter and May in London and Paris; for the UK government's legal service, in the Department for Environment, Food and Rural Affairs (Defra); and with NGOs Friends of the Irish Environment and An Taisce, the National Trust for Ireland, where he was Natural Environment Officer and in-house Solicitor. Andrew is a practising lawyer and has been involved in public interest environmental litigation for many years, including before the Irish, English and EU courts. He acted for the successful litigant in the landmark "Climate Case Ireland" before the Supreme Court of Ireland in 2020. Andrew is a graduate of Oxford University (BA, Law) and Cambridge University (LLM). He also holds an MSc in Biodiversity and Conservation from Trinity College Dublin (TCD) and a PhD from TCD.

Gregory Jones KC, BL practises from Francis Taylor Building, Temple, London and also from the Bar Library, Belfast and the Law Library, Dublin. Educated at New College, Oxford and University College, London, Gregory's expertise includes town and country planning, Parliamentary, environmental, EU, compulsory purchase, administrative and local government law. Gregory is the Chancellor of the Diocese of Manchester and a Bencher of both Lincoln's Inn and Inner Temple. He is a Fellow of the Centre of European Law, King's College, London and a Distinguished Visiting Fellow of St. Stephen's House, Oxford University. In 2015 Gregory was appointed by the Irish government to chair an independent review of An Bord Pleanála reporting in February 2016. He is an elected Alderman in the City of London and a founding director of a renewable energy company.

José Vicente López-Bao is a conservation scientist at the Spanish National Research Council (CSIC), Spain. He is interested in the integration of quantitative, interdisciplinary and transdisciplinary approaches to face conservation challenges, with particular interest in nature conservation, human-wildlife interactions, and large carnivores and other megafauna in human-dominated landscapes.

Micheál Ó Briain worked as an official in the Nature Unit of DG Environment in the European Commission from 1992 to 2021, spending the last ten years as its Deputy Head of Unit. He worked on a range of nature and biodiversity policy issues including overseeing the implementation of the EU Birds Directive and co-ordinating the implementation of the EU Biodiversity Action Plan. Micheál played an active role in the establishment of the Natura 2000 network under the Habitats Directive including co-ordinating work on its financing and management. He co-ordinated the Fitness Check evaluation of the Birds and Habitats Directive between 2014 and 2016 and played a key role in the development of the follow up "Action Plan for Nature People and the Economy". Prior to working in Brussels Micheál was national director of the Irish Wildbird Conservancy, now called BirdWatch Ireland, the Irish partner of BirdLife International. He is a graduate and former research associate of University College Dublin where he completed a PhD in Zoology.

Hendrik Schoukens is a Professor of Environmental Law at Ghent University, and has published on a wide array of topics such as nature protection, ecological restoration, climate change, access to justice, sustainability and environmental impact assessment. He is also a solicitor at the Bar of Brussels, in which capacity he frequently litigates high-profile environmental cases. Hendrik is also an elected city councillor for environmental and planning matters in the Belgian municipality of Lennik.

Christoph Sobotta is a référendaire (legal secretary) in the Chambers of Advocate General Juliane Kokott at the Court of Justice of the European Union. Since 2003, he has assisted the Advocate General with the drafting of more than 180 opinions, in particular in the area of environmental law. In 2000 and 2001, he assisted her predecessor Siegbert Alber. Between 2001 and 2003, he was a case handler in the infringement unit of DG Environment of the European Commission. Christoph studied law at the University of Frankfurt am Main and at Keele University in the UK. In his doctoral thesis, he discussed Transparency in the European Union. He has published and spoken on many issues of EU law, in particular on EU environmental, data protection and constitutional law as well as on the Court of Justice.

John Temple Lang was an Irish solicitor, an Adjunct Professor at Trinity College Dublin and a Senior Visiting Research Fellow at Oxford. He worked in the Legal Service of the European Commission and as a Director in the Competition DG. In 2000 he went back into private practice with Cleary Gottlieb Steen & Hamilton LLP. He was awarded honorary doctorates from Trinity College Dublin and

Lund University; he published a book on EU law and more than 350 papers; and he was responsible for drafting the Birds Directive and was involved in drafting the Habitats Directive. Dr. Temple Lang sadly passed away in November 2022, and this book is dedicated to his memory, in reflection of his outstanding contribution to nature conservation.

ACKNOWLEDGEMENTS

The editor would like to thank the Sutherland School of Law at University College Dublin for hosting the conference that led to this book, and for the research sabbatical that enabled the editor to focus on the task at hand. Special thanks are due to Hannah Ferguson, Katie Stokes, John Baddeley, Amy Johnston and all at Routledge, who were unfailingly patient in allowing sufficient time for the completion of this project. Special thanks are also due to Micheál Ó Briain who, in addition to contributing an excellent chapter, kindly provided the cover image for this book, which shows Achill Island in Ireland.

1

INTRODUCTION

Andrew L. R. Jackson

This edited collection began life as a conference held at University College Dublin in June 2018, under the title "Key challenges in European nature law and policy". Some of those who spoke that day have kindly contributed chapters here, while others who did not participate in the event were later approached to contribute. The conference was held the day after the launch of a monograph published by the editor of this collection, entitled "Conserving Europe's Wildlife: Law and Policy of the Natura 2000 Network of Protected Areas",[1] also published by Routledge.

That book traces the origins and negotiation of the EU's Birds and Habitats Directives, and its launch was therefore an opportunity to bring together some of the key figures in that history, including for example John Temple Lang – who drafted the Birds Directive and was involved in the drafting of the Habitats Directive – as well as Stanley Johnson, who as a Commission official in what is now DG Environment played an important role in launching the Habitats Directive project. The conference also brought together many of those who have played and continue to play an integral role in the life of the Nature Directives since their adoption. Buoyed by the success of the event, the editor decided to pursue the idea of publishing the contributions, as well as commissioning a few others specifically to contribute to the project, and four years on you hold in your hands (or you see on your screen, eBook readers) the outcome!

The idea behind the conference and this book was to bring together contributions from the European Union and the UK (which, at the time the conference was held, had committed to Brexit but had not yet Brexited) in order to record and assess the state of nature law and policy at that moment in time, before the UK began to tread its own path. The interruption caused by the Covid-19 pandemic extended the delivery of this collection over a longer period than would

DOI: 10.4324/9780429299100-1

otherwise have been the case, but this has allowed in consequence for a depth of analysis in the chapters that otherwise would not have been possible. The contributions now capture not just a snapshot in time, as it were, but more of a moving picture.

During the production of this book, two major figures in the history of European nature conservation sadly passed away: Claus Stuffmann in November 2020 and John Temple Lang (author of Chapter 3 of this collection) in November 2022. As Temple Lang himself commented in 2013, "Without [Claus Stuffmann, the European Commission's senior negotiator of the Birds and Habitats Directives] I doubt if either [directive] would have been adopted. He patiently and professionally steered the ships through many shoals … It was an achievement of the best kind that Commission officials can do."[2] Stuffmann headed the unit responsible for nature conservation in what would ultimately become DG Environment of the European Commission from 1976 to 1994. Before that, Stuffmann was a member of Commission President Jean Rey's cabinet, before joining the cabinet of Altiero Spinelli, then Commissioner for Industrial Affairs, Science and Technology, and one of the founding figures of the EU and of EU environmental policy.[3] Member State negotiators of the Habitats Directive variously described Claus Stuffmann as "a wonderful person", as someone who should be "awarded a prize for European diplomacy", and as someone who "deserves a MEDAL" from Europe.[4] For his part, in a long and distinguished career that encompassed both competition law and environmental law,[5] John Temple Lang *inter alia* drafted the Birds Directive and co-drafted the Habitats Directive whilst working for the Commission's legal service. In the aftermath of the Birds Directive's adoption, Temple Lang advised stakeholders as to the new possibilities for nature opened up by the Directive, which led ultimately to the Important Bird Area system that now applies globally.[6] Temple Lang described his idea for the Birds Directive as "extremely simple": namely, "to extend the [criteria for identifying areas of international importance] to all migratory species, and then in parallel to produce a comprehensive list of every place in Europe that met those criteria, and argue that there was a legal obligation to protect them all. This idea succeeded, even better than I had hoped".[7] A passionate ornithologist throughout his life, Temple Lang was a co-founder of the Irish Wildbird Conservancy in the 1970s – now BirdWatch Ireland – and was an active member of various other environmental organisations, including the European section of BirdLife International.[8] Together, Claus Stuffmann and John Temple Lang played key roles in laying the groundwork for the Natura 2000 network and the system of nature conservation that we know today in Europe.

On the 20th anniversary of the Birds Directive's adoption, Alistair Gammell, formerly of the Royal Society for the Protection of Birds, commented that "[Tomorrow's Europeans] will be grateful that a handful of people in the early 1970s had the foresight to act to try to protect these areas and species and that a movement of European citizens … fought, in some cases successfully, to conserve them … Tomorrow's Europeans will thank us for that."[9] In reflection of their outstanding contributions to nature conservation, the present book is dedicated to the memories of Claus Stuffmann and John Temple Lang.

Following the present introductory chapter, in Chapter 2 (Recent developments in the implementation of EU nature legislation) Micheál Ó Briain discusses recent developments in the implementation of EU nature legislation, including challenges to progress in implementation since 1992; the EU fitness check evaluation of the Birds and Habitats Directives, including the key findings of that fitness check; and the subsequent EU Action Plan for Nature, People and the Economy. The chapter concludes by considering the role of the Nature Directives and Natura 2000 in post-2020 EU biodiversity policy.

In Chapter 3 (The requirements for an effective international nature conservation system, and the European Union Birds and Habitats Directives), John Temple Lang considers the requirements for an effective international nature conservation system, then discusses the EU's Birds and Habitats Directives in the light of these requirements. The chapter concludes with comments by way of an overall assessment of EU nature conservation policy.

In Chapter 4 (Ensuring compliance: regulatory regimes under the EU Nature Directives), Liam Cashman explores how national authorities can ensure compliance with site conservation obligations placed on landowners, land managers and land users. The chapter considers the concepts of 'enforcement' and 'environmental compliance assurance'. It then touches on the underlying logic of the Birds and Habitats Directives and the range of obligations to which they give rise, including obligations under regulatory regimes applicable to land owners. The chapter outlines how compliance problems and challenges can be characterised, the principles of compliance assurance, the range of interventions compliance assurance covers, and strategic approaches to using them. The chapter finishes by looking at Ireland as a case study, including for example the experience with Irish raised bogs.

In Chapter 5 (The margin of appreciation when applying Article 6(3) and (4) of the Habitats Directive), Christoph Sobotta begins by describing the nature of the decisions that need to be adopted in respect of Article 6(3) and (4) of the Habitats Directive (the 'appropriate assessment' and derogation provisions in respect of Natura 2000 sites), in particular the complexity of the appraisals that are required. From a legal perspective, the standard of judicial review to be applied to these decisions is one of the most important topics in the area. The chapter develops a possible standard of judicial review by drawing inspiration from the jurisprudence of the Court of Justice of the European Union (CJEU) with regard to complex appraisals by EU institutions. Subsequently, the chapter analyses the jurisprudence of the CJEU on the application of the Habitats Directive, and in particular Article 6(3), in the light of the identified standard of review. On close examination, Sobotta argues, the strict jurisprudence on Article 6(3) is an expression of the stricter enforcement of procedural requirements. On the other hand, the derogation laid down in Article 6(4) should, he argues, provide for a wider margin of appreciation.

In Chapter 6 (The trigger for Article 6(3) appropriate assessment: the gateway revisited), Gregory Jones KC, BL examines the approach taken at the so-called 'screening' stage of the appropriate assessment process in the *Hart*[10] line of authority, which relates to whether or not avoidance or mitigation measures can be taken into account in determining whether a plan or project is likely to have significant effects on a Natura 2000 site. Jones argues that a failure properly to understand the differences in the relationship between environmental assessment (under the EIA or SEA Directives) and appropriate assessment (under the Habitats Directive) led some of the judiciary in the UK and Ireland astray in the *Hart* line of authorities. Focusing on the UK and Ireland, the chapter also asks whether the judgment of the CJEU in *People Over Wind*[11] has now settled the matter. Finally, the chapter considers whether the *Hart* approach to the Habitats Directive may be indicative of a wider difference of common law judicial culture that might lead in time to a divergence between the UK and EU, even where the articles of the Directive may have been preserved for the time being in UK law by being 'cut and pasted' into domestic law.

In Chapter 7 (Supporting nature conservation through an ecosystem services policy), Ian Hodge outlines the way in which a British Ecosystem Services Policy (BESP) might operate at a national scale to replace the role of the Common Agricultural Policy (CAP). Hodge argues that nature conservation policy needs to respond in particular to the challenges of climate change, to do more to support nature conservation in the wider countryside, and to integrate nature conservation measures better with other rural land values and practices. This indicates, he argues, a need to move further beyond the site-based policy that has dominated in the past. However, the UK lacks, the chapter argues, the systematic governance arrangements through which to implement a new approach. While the principle of a long-term ambition has been established through the UK's 25 Year Environment Plan, parallel changes in governance arrangements are needed that can bring it to fruition. Amongst other things, the chapter discusses the way in which a BESP could contribute towards the development of new nature conservation policies.

In Chapter 8 (Judicial Review and nature cases in the UK: current challenges and future opportunities), Carol Day considers the use of judicial review in nature cases in the UK. While the success rates associated with judicial review are low, the mechanism nevertheless remains a vital backstop for checking the abuse of public power, Day argues. The chapter reviews some of the ways in which judicial review, and other participatory rights, need to be strengthened – including as regards the intensity of review, legal costs and prohibitive expense, and remedies – and considers the strategically important role that judicial review can play in environmental protection. Amongst other things, the chapter considers the requirements of the Aarhus Convention in this area, environmental governance arrangements in the UK post-Brexit, and the potential need for a specialist Environmental Court or Tribunal.

In Chapter 9 (Science and the interpretation of the EU Habitats Directive by the Court of Justice of the EU: 2+2=5?), Hendrik Schoukens reviews the use of scientific data in nature litigation and also the use of the precautionary principle, when precise and conclusive scientific data are not available. In refusing permits for plans or projects or derogation applications on the basis of the precautionary principle, the CJEU has turned the Habitats Directive into a highly effective legal instrument, Schoukens argues. However, this is premised on the idea that science is always rational and objective. Rather than using the precautionary principle as a 'shield' against scientific controversies, Schoukens argues that the EU courts may sooner or later need to call in the help of scientific experts, particularly in light of the increasingly complex and contested legal and ecological questions arising.

In Chapter 10 (Strict protection of species in the EU: controversies and trends), Yaffa Epstein, José Vicente López-Bao and Guillaume Chapron discuss the requirements for the protection of species under Articles 12 to 16 and Annexes IV and V of the Habitats Directive, trends in the interpretation and enforcement of these requirements, and ongoing open questions. Amongst other things, the authors discuss the CJEU's leading decisions in Cases C-674/17 *Tapiola* (wolf hunting in Finland),[12] C-441/17 *Commission v Poland* (Białowieża Forest),[13] and Joined Cases C-473/19 and C-474/19 *Skydda Skogen* (strict protection of species in Sweden).[14]

In the concluding Chapter 11 (Conservation of areas outside the Natura 2000 network: an historical perspective with an eye to the future), the editor of the present collection returns to the archives to consider the negotiation and agreement of the parts of the Habitats and Birds Directives that relate to the protection of areas outside Natura 2000. By understanding how the relevant provisions of these Directives came to be agreed, and by understanding what was discounted or excluded during negotiations, as well as the positions adopted by different parties, we may perhaps arrive at a clearer understanding of the arguments that have in the past framed and constrained the development of legal protections for areas outside the Natura 2000 network. Similar issues may or may not arise in future discussions regarding the conservation of such areas – for example, during the forthcoming negotiations of the proposed EU Nature Restoration Law – but in understanding the policy dynamics that were at play and the discussions that took place during earlier periods in EU environmental policy, today's policy makers and stakeholders may perhaps be better positioned to secure a positive outcome for biodiversity when the next occasion arises.

As is well known, despite some positive stories and successes, the current outlook for the biodiversity and climate crisis globally is not good. Recent reports of the Intergovernmental Panel on Climate Change (IPCC) and the Intergovernmental Science-Policy Platform on Biodiversity and Ecosystem Services (IPBES) make for increasingly grim reading. The IPCC's WGII report of the Sixth Assessment Report recently concluded with *very high confidence* that "Any further delay in concerted anticipatory global action on adaptation and mitigation will miss a brief and rapidly closing window of opportunity to secure a liveable and sustainable future for all."[15] For its part, IPBES's global report in 2019 concluded that "Goals for conserving and sustainably using nature and achieving sustainability cannot be met by current trajectories, and goals for 2030 and beyond may only be achieved through transformative changes across economic, social, political and technological factors."[16] 'Transformative change' is defined in a footnote (incongruously, given the gravity of the statement) as "A fundamental, system-wide reorganization across technological, economic and social factors, including paradigms, goals and values."[17]

Interestingly, we need not look too far into the distant EU past to find transformative change of precisely this type dominating the headlines. Back in February 1972, Sicco Mansholt of the Netherlands – architect of the common agricultural policy as Agriculture Commissioner from 1958 to 1972 – wrote a lengthy letter to then Commission President Malfatti.[18] Mansholt was very interested in environmental and development issues and had been profoundly affected by the Club of Rome's ideas and by his friend Professor Jan Tinbergen, winner of the Nobel Prize for Economics in 1969.[19] In his letter to Malfatti, Mansholt called for an immediate U-turn in the EU's economic policy: an abandonment of economic growth maximisation and the development of a "clean and recycling" economy, with GNP to be replaced by gross national happiness as a measure of progress. To many people's surprise, in March 1972 - just a month after sending his letter – Mansholt was made interim Commission President.[20] In this capacity, he represented the EU at the famous UN Stockholm Conference of 1972 and repeated his *Limits to Growth*-inspired ideas on the international stage.[21] His speech was said to have been acclaimed by young demonstrators.[22] Given current debates and discourses – from planetary boundaries[23] to degrowth[24] to doughnut economics[25] – perhaps Mansholt will not be the last Commission president to champion such ideas.

The EU Biodiversity Strategy for 2030[26] echoes IPBES's language of transformative change, and the commitments made thereunder are undoubtedly ambitious. Delivering on the strategy by, for example, greatly expanding the EU's protected area network, establishing a Trans-European Nature Network, and agreeing in due course an ambitious EU Nature Restoration Law[27] would certainly transform the European landscape anew. In the UK (or in England, at least), things appear to be moving in rather a different direction, with plans recently announced to introduce 'investment zones' in England with "liberalised planning rules" under which the Government will "remove burdensome EU requirements which create paperwork and stall development but do not necessarily protect the environment".[28] This, combined with the announcement of the new Retained EU Law (Revocation and Reform) Bill 2022, which proposes sunsetting the majority of retained EU law so that it expires on 31 December 2023,[29] has left many fearing for the future of the UK's nature conservation regime.

Pending future developments in the EU and the UK, it is hoped that the current collection contributes usefully by way of explaining, discussing and critiquing nature law and policy in Europe, on the brink of transformative change.

Notes

1 Andrew L.R. Jackson, *Conserving Europe's Wildlife: Law and Policy of the Natura 2000 Network of Protected Areas* (Routledge 2018).
2 Ibid, p. 109.
3 Ibid, pp. 41–42. Also see: Delegation of the Commission of the European Communities (Information service), 'A biographical note: Claus Stuffmann, Staff member of the cabinet of Altiero Spinelli', March 1975, available at http://aei.pitt.edu/78334/1/Stuffmann.pdf.
4 Jackson, supra n 1, at footnote 2 on p. 150.
5 See Obituaries, 'John Temple Lang Obituary: Savvy Lawyer Dedicated to Public Service', *The Irish Times*, 19 November 2022.
6 Jackson, supra n 1.
7 Ibid, p. 42.
8 Supra n 5.
9 Alistair Gammell, 'The contribution of BirdLife International to the Birds Directive and the management of birds and their habitats in Europe. 20 years with the EC Birds Directive', Proceedings from a conference on the Council's Directive on the Conservation of Wild Birds, 18–19 November 1999 Elsinore, Denmark (the Danish National Forest and Nature Agency 1999).
10 *R (Hart DC) v SSCLG* [2008] EWHC 1204.
11 Case C-323/17 *People Over Wind and Sweetman v Coillte Teoranta*, EU:C:2018:244.
12 Case C-674/17 *Luonnonsuojeluyhdistys Tapiola Pohjois-Savo — Kainuu ry,* EU:C:2019:851.
13 Case C-441/17 *Commission v Poland,* EU:C:2018:255.
14 Joined cases C-473/19 and C-474/19 *Föreningen Skydda Skogen and Others v Länsstyrelsen i Västra Götalands län and Others,* EU:C:2021:166.
15 IPCC, 'Summary for Policymakers', in H.-O. Pörtner et al (eds.), *Climate Change 2022: Impacts, Adaptation and Vulnerability. Contribution of Working Group II to the Sixth Assessment Report of the Intergovernmental Panel on Climate Change* (Cambridge University Press) pp. 3–33, doi:10.1017/9781009325844.001, at D.5.3.
16 IPBES, Summary for policymakers of the global assessment report on biodiversity and ecosystem services of the Intergovernmental Science-Policy Platform on Biodiversity and Ecosystem Services. S. Díaz et al (eds.). *IPBES Secretariat*, Bonn, Germany, at p.14.
17 Ibid, at p.14, footnote 4.
18 Sicco Mansholt, Letter from Sicco Mansholt to Franco Maria Malfatti, President of the European Commission, dated 14 February 1972, the Dutch original http://www.cvce.eu/viewer/-/content/51303966-0532-46bc-89c7–271ef294eb13/en and an English translation https://themansholtletter.hetnieuweinstituut.nl/sites/default/files/brief_mansholt_malfatti_en1.pdf
19 For more, see Jackson, supra n 1.
20 Ibid.
21 Ibid. Also see the candid "negative growth" remarks Mansholt made shortly after leaving the Commission: see Vanya Walker-Leigh, 'Interview with Sicco Mansholt' (August/September 1974) 4 *The Ecologist* 259–260.
22 See Jan van der Harst, 'Sicco Mansholt: Courage and Conviction', in Michel Dumoulin (ed.), *The European Commission, 1958–1972: History and Memories (Deluxe edition).* Updated 2014. Office for Official Publications of the European Communities.

23 Will Steffen et al, 'Planetary Boundaries: Guiding Human Development on a Changing Planet' (2015) 347(6223) *Science* 347–357.
24 Federico Demaria et al, 'What Is Degrowth? From an Activist Slogan to a Social Movement' (2013) 22 *Environmental Values* 191–215.
25 Kate Raworth, *Doughnut Economics: Seven Ways to Think Like a 21st-century Economist* (Random House 2017).
26 European Commission, 'EU Biodiversity Strategy for 2030: Bringing Nature Back into Our Lives' COM(2020) 380 final, Brussels, 20.5.2020.
27 European Commission, 'Proposal for a Regulation of the European Parliament and of the Council on Nature Restoration' COM(2022) 304 final, 22.6.2022.
28 HM Government, Guidance: Investment Zones in England, published 24 September 2022, https://www.gov.uk/government/publications/investment-zones-in-england/investment-zones-in-england
29 HM Government, News story: The Retained EU Law (Revocation and Reform) Bill 2022, published 22 September 2022.

2

RECENT DEVELOPMENTS IN THE IMPLEMENTATION OF EU NATURE LEGISLATION

Micheál Ó Briain[1]

Introduction

Europeans live in one of the most densely populated regions of the world, with a long history of land use and development that has greatly shaped the biodiversity of our continent. Some of this interaction with nature has been relatively benign, leading to the creation of rich and diverse cultural landscapes. However, there has been large-scale loss and degradation of nature in Europe, particularly over the course of the 20th century. A study on birds for the European Commission in 1977 revealed a dramatic decline in many species during the century, with populations of groups such as birds of prey greatly depleted.[2] Between 1900 and the mid-1980s two-thirds of wetlands in Western Europe were estimated to have been lost.[3] Similar large-scale losses of high-nature-value farmland, sand dunes and heaths had also occurred. The key drivers of these losses were land use changes, urban and infrastructure development and pollution.

International legal and policy responses to the decline of species have largely been put in place since the 1970s. Much of the initial focus was on wildlife legislation, particularly concerning threatened species. Increasingly the protection of nature is also based on wider environmental grounds. Healthy terrestrial and marine ecosystems play a vital role in regulating the environment and are now being recognised as essential in climate change adaptation and mitigation. Ensuring the health of nature is also an economic imperative as ecosystem goods and services, such as water purification and provision, insect pollination, preserving landscape and amenity values, and supporting tourism and recreation, deliver a whole range of direct and indirect economic benefits that sustain jobs and growth, particularly in rural areas.

The Birds Directive and Habitats Directive (frequently referred to as the Nature Directives) are the key EU legal instruments for the conservation and

DOI: 10.4324/9780429299100-2

sustainable use of nature. They provide a common framework for the protection of species and habitats of EU conservation concern and are the cornerstone of EU biodiversity policy. The Directives are central to fulfilling EU commitments under international conventions and agreements, such as the Convention on Biological Diversity, the Bern Convention on European Wildlife, the Convention on Migratory Species, and the Agreement on the Conservation of African-Eurasian Migratory Waterbirds. Effective implementation of the Nature Directives is critical to achieving the overarching objective of the new EU Biodiversity Strategy.[4] This aims to put Europe's biodiversity on the path to recovery by 2030 for the benefit of people, the planet, the climate and our economy, in line with the 2030 Agenda for Sustainable Development and with the objectives of the Paris Agreement on Climate Change.

Adopted in 1979, the Birds Directive provides a framework for the conservation of all species of naturally occurring birds in the wild state in the European territory of the EU Member States. Its overall objective is to maintain the populations of all wild bird species at a level that corresponds to their ecological, scientific and cultural requirements, or to adapt the population of these species to that level. The Habitats Directive, adopted 13 years later in 1992, extended the species coverage to more than 1,200 rare, threatened or endemic species of wild animals and plants, collectively referred to as species of Community interest. A particular innovation was the extension of protection to habitat types of EU conservation concern, now covering 231 rare habitat types of Community interest. Its overall objective is to maintain or restore these habitats and species to favourable conservation status.[5]

Both Directives apply similar mechanisms. They require a sufficient number and area of the most important sites to be protected and managed in a coherent network called Natura 2000. This comprises Special Protection Areas (SPAs) designated under the Birds Directive and Special Areas of Conservation (SACs) designated under the Habitats Directive. It is established for species and habitat types listed in Annex I and II of the Habitats Directive and in Annex I of the Birds Directive, as well as for migratory birds. They also require and encourage Member States to take habitat conservation measures beyond the Natura 2000 network. Furthermore, Member States must establish a strict protection regime for all naturally occurring wild bird species and for species listed in Annex IV of the Habitats Directive, both inside and outside Natura 2000 sites. Whereas certain bird species listed in Annex II of the Birds Directive may be hunted and other species listed in Annex V of the Habitats Directive may be exploited Member States must ensure that this is sustainable and does not jeopardise conservation efforts. Finally, Member States must also ensure adequate knowledge, data availability and awareness to underpin implementation of the Directives.

The Natura 2000 network is the most important mechanism to achieve the overall conservation objective of the Directives. Member States are required to contribute to this network in relation to their importance for protected species and habitats. Site selection is made exclusively on scientific grounds. For SPAs

Member States identify and designate the most suitable territories on the basis of objectively verifiable ornithological criteria. The knowledge on birds is much better than for most other taxonomic groups and scientific references (Important Bird Areas) have been developed since the early days of implementation of the Birds Directive. Although not legally binding they are of high value in helping identify SPAs and in assessing the completeness of this component of the Natura 2000 network (Case C-3/96 *Commission v the Netherlands*).

Selection of SACs is a more complex process. Member States first identify and propose important locations for the protected species and habitats present on their territory. The European Commission then selects, with the assistance of the European Environment Agency's (EEA) Topic Centre for Biodiversity and scientific experts, and in agreement with the Member States, Sites of Community Importance (SCIs) within the framework of each of nine biogeographical regions covered by the Habitats Directive. Once selected, by way of Commission Decisions, the SCIs become part of the Natura 2000 network. Member States then have up to six years to designate them as SACs and to establish the necessary conservation measures to achieve the site conservation objectives.

Although it does include nature reserves and national parks, Natura 2000 is based on a much wider concept of nature conservation in which socio-economic activities that are not damaging to the nature values are allowed. The majority of sites in Natura 2000 are under some form of active land use and constitute an integral part of the wider countryside. A broad range of activities are carried out in Natura 2000 sites across Europe, including farming, forestry, fisheries and tourism. These activities are important for local economies, especially in more rural and remote regions.

Whereas it is not an obligation under the Directives, the development of management plans is strongly advocated by the European Commission as a tool for establishing the necessary conservation measures. This also allows for greater transparency of process and stakeholder engagement. It also provides opportunities to identify and optimise the multiple benefits that can accrue from site management.

New developments that present a risk to the conservation objectives of Natura 2000 are not automatically excluded but must be subject to appropriate assessment. Damaging developments may only proceed under certain conditions, set out in Article 6(4) of the Habitats Directive.

Challenges to progress in implementation since 1992

Implementation of the Nature Directives has taken place in the context of successive enlargements of the EU, which have required some adaptations to the Annexes of the Directives as well as the addition of new biogeographical regions. Otherwise, this legal framework has remained largely consistent and stable since 1992.

There have been significant challenges that have delayed implementation and impeded achievement of the objectives of the Nature Directives, both in relation to site and species protection. The issue of hunting in particular proved to be one of the more difficult and polarising issues. Before the adoption of the Birds Directive hunting of wild birds appears to have been generally more extensive, with more species hunted, longer hunting seasons, extending in some countries significantly into the Spring migration period. A judgment of the EU Court of Justice in 1994[6] confirmed the need for 'complete protection' during the pre-nuptial migration. A subsequent Commission-proposed modification of the Directive to allow for more flexibility on hunting provisions[7] was rejected by the European Parliament and was ultimately withdrawn.

In response to continued tensions on hunting, the Commission launched a sustainable hunting initiative in 2001 to promote dialogue and build trust between the European Federation for Hunting and Conservation (FACE) and BirdLife International, the main stakeholder groups involved in this debate. This involved developing a programme of scientific, conservation and awareness raising measures to improve understanding of the legal and technical aspects of the Birds Directive's provisions on hunting. It included development by the Commission of guidance on sustainable hunting under the Birds Directive[8] as well as the compilation of scientific information relevant to defining hunting seasons under the Directive. The dialogue led to the signing of an Agreement on sustainable hunting between BirdLife and FACE in 2004,[9] a framework of cooperation between these organisations that has continued to the present day, even if relations at national level, and sometimes even at EU level, are not always successful.

There was also significant opposition to Natura 2000 designations in different Member States that contributed to serious delays in the establishment of the network, with the timeframes foreseen under the Directives not being met. Funding was provided under the EU LIFE programme to assist Member States in identification of sites, both for terrestrial and marine environments, which significantly strengthened the knowledge base. However, as part of its response to these delays, the Commission also took enforcement action against Member States. This resulted in a series of rulings of the Court of Justice of the EU under both Directives relating to failures to classify the most suitable territories as SPAs[10] and to propose complete lists of SCIs.[11] Furthermore, the Commission also threatened to block the use of funding under Cohesion Policy in cases where Member States had failed to propose their Natura 2000 sites and there was a decision to suspend funding temporarily for the Walloon Region of Belgium.[12]

Following these actions, the number of Natura 2000 sites increased substantially from the late 1990s. The Natura 2000 network in 2019 was comprised of 27,852 sites covering 1,358,125 km^2 (about 18% of EU land and 9% of EU seas).[13] Even with a reduction in the number of sites from January 2021, when the Nature Directives no longer apply to the United Kingdom,[14] this represents the largest multinational co-ordinated protected area network in the world. It is almost complete on land but there are some remaining gaps for the marine area, particularly for the offshore marine environment.

Despite this progress, there have been continuing challenges and conflicts related to implementation of both the site and species protection provisions of the Directives. This has frequently related to the application of the protection and procedural safeguards applying to Natura 2000 sites under Article 6(2) and 6(3) of the Habitats Directive. Some of the conflicts related to the protection and management of species have been linked to populations of a small number of species such as Wolves, Beavers and Cormorants that are undergoing recovery. Their return to areas where they may have been absent for generations has led to significant challenges of co-existence with certain stakeholder groups.

Tensions have also been evident at high political level in some Member States. For example, in 2009 the then Dutch Prime Minister Jan-Peter Balkenende wrote to Commission President José Manuel Barroso seeking a possible relaxation of the protection of Natura 2000 areas, arguing that this was needed to strike a balance between ecological value, economic interests and other uses. Likewise, in 2011, the then Chancellor of the Exchequer in the United Kingdom, George Osborne, expressed concerns about 'gold plating' of EU environmental rules in areas such as the Habitats Directive, fearing that this that may place 'ridiculous costs' on business in the United Kingdom. This triggered a review of implementation of the Habitats Directive in England and Wales aimed at reducing burdens on business while maintaining the integrity of the purpose of the Directive.[15]

The EU fitness check evaluation

In addition to this review of implementation at national level in the United Kingdom, at EU level a detailed 'fitness check' was carried out, representing the most comprehensive evaluation ever undertaken of the performance of the Nature Directives. It did not arise from the formal reporting requirements of the Birds and Habitats Directives. However, by providing up-to-date information on the conservation status and trends of species and habitats protected by the Nature Directives the Commission and EEA's 2015 'State of Nature in the EU' assessment was a particularly valuable input to the evaluation.[16]

The evaluation was carried out within the framework of the European Commission's Regulatory Fitness and Performance programme (REFIT), an initiative aimed at making EU law simpler, more efficient and effective, seeking to reduce any unnecessary regulatory costs, thus contributing to a clear, stable and predictable regulatory framework supporting growth and jobs. The emergence of the fitness checks was in line with an approach that had been initiated under the Barroso-led Commissions, initially focusing on the role of impact assessment for new legislative proposals and latterly on evaluation of existing legislation. Within this framework, different pieces of the environmental acquis have been and are being subject to evaluations.

A decision to subject the Nature Directives to a Fitness Check was announced in October 2013 in recognition that these pieces of legislation had been in place for a number of years and that it was timely to evaluate their impact.[17] It aimed to provide an evidence-based critical analysis of whether EU actions are proportionate to their objectives and delivering as expected. It examined their performance against five standard criteria:

- Effectiveness (Have the objectives been met?);
- Efficiency (Were the costs involved reasonable?);
- Coherence (Does the policy complement other actions or are there contradictions?);
- Relevance (Is EU action still necessary?);
- EU added value (Can or could similar changes have been achieved at national/regional level, or did EU action provide clear added value?).

The evaluation involved examining implementation and integration successes and problems, the costs of implementation and non-implementation of the legislation, the administrative burden of implementation, and the opportunities to reduce it without compromising the integrity of the purpose of the Directives. It considered what has worked well or poorly, and compared actual performance to earlier expectations.

A Steering Group of Commission Services oversaw the process and agreed the mandate for the Fitness Check, which was published in early 2014.[18] In September 2014 President Jean-Claude Juncker's Mission Letter to incoming Environment, Maritime Affairs and Fisheries Commissioner Karmenu Vella, asked him to "*carry out an in-depth evaluation of the Birds and Habitats directives and assess the potential for merging them into a more modern piece of legislation*" as one of the priority actions for his mandate. This considerably raised suspicions among environmental NGOs about the overall intent of the evaluation and was probably a very significant motivational factor for the 'Nature Alert Campaign', which they launched in response to the fitness check.

The Commission carried out a very extensive and transparent evidence-gathering process, mainly between November 2014 and November 2015, supported by a 15-month study contract.[19] This included questionnaires sent to public authorities and stakeholder groups in all EU Member States, not only involving the competent authorities for nature and environmental NGOs but also other public and private bodies that were affected by the Directives. With a view to examining challenges of implementation in greater detail, the Commission also undertook fact-finding visits to ten representative Member States. There was also an extensive literature search, gathering over 1,800 relevant Commission studies, published and peer-reviewed papers and other key documents.[20]

An online public consultation attracted a record number of responses (552,472), boosted by the 'Nature Alert' campaign run by environmental NGOs who opposed any changes to the Directives. Their campaign contributed close to 500,000 of the responses. However, the results of the public consultation also revealed that there were other smaller campaigns from different stakeholder groups, particularly from agriculture and forestry groups.[21] A conference was held in November 2015 to discuss the emerging findings of the evaluation.

During the period of the evaluation, other EU institutions underlined the importance of the Nature Directives. In their response to the Commission Communication on the mid-term review of the EU Biodiversity Strategy, the follow-up Conclusions of the Council, the Resolution of the European Parliament and the Opinion of the European Economic and Social Committee all emphasised the central role of the Nature Directives in achieving EU biodiversity targets.

The draft Commission Staff Working Document, presenting the analysis and findings of the fitness check, was subject to examination by the Commission's Regulatory Scrutiny Board in April 2016. Following further Commission services consultation and an orientation debate in the College of Commissioners on 7 December 2016, the Fitness Check conclusions were published in the final Commission Staff Working Document on 16 December 2016.[22]

Key findings of the fitness check

The overall headline conclusion of the nature fitness check was that

> Within the framework of broader biodiversity policy the Nature Directives are fit for purpose but fully achieving their objectives and realising their full potential will depend on substantial improvement in their implementation in relation to both effectiveness and efficiency, working in partnership with different stakeholder communities in the Member States and across the EU, to deliver practical results on the ground.

This conclusion, which effectively meant that the Directives would not be subject to any proposed legal amendment by the then Commission, was based on detailed assessment of each of the five evaluation criteria, responding to the questions set out in the fitness check mandate.

Effectiveness

The evaluation confirmed that the general objectives of the Directives had not yet been met and it could not predict when they would be fully achieved. Despite continuing declines of many species and habitats, the evaluation determined that the status and trends of bird species as well as other species and habitats protected by the Directives would be significantly worse in the absence of the Directives. Furthermore, there was good evidence that improvements in the status of species and habitats were taking place where targeted actions were happening at a sufficient scale. The terrestrial part of the Natura 2000 protected areas network was now largely established although gaps remained for the marine environment. National systems of species protection and sustainable use were in place. Progress on habitat conservation and delivery of ecosystem services beyond Natura 2000 was more limited.

Increased funding for nature had been made available as a result of the Directives, although it was still insufficient and there were problems relating to the ability to uptake funds. The LIFE programme was strategically important but integration with agriculture and other EU funds had been less successful. For the 2014–2020 funding period, the Commission had promoted the development of Prioritised Action Frameworks (PAFs) by Member States to strategically identify needs and priorities for investment in Natura 2000.[23] These planning tools were aimed at strengthening the integration of Natura 2000 financing into the use of relevant EU financial instruments. However, in practice, their impact had been constrained by their varying quality and by the extent to which they had been taken into account by Member States in their use of the key sectoral funds.

The effectiveness of the Directives had been significantly hindered by delays in establishment of Natura 2000 and in putting in place the necessary conservation measures. It was estimated that only 50% of the Natura 2000 sites had management plans. Key factors behind the shortcomings in implementation included limited financial and human resources, poor enforcement, weak integration of nature objectives into other policy sectors, insufficient knowledge and access to data as well as failures in communication and stakeholder engagement.

Several of these problems were also subsequently highlighted in a separate audit of the performance of Natura 2000 by the European Court of Auditors, published in 2017.[24]

Efficiency

The evaluation concluded that the multiple benefits of the Directives far exceed the costs, although there was still insufficient recognition of these benefits in public policy. The benefits of Natura 2000 have been estimated at €200–300 billion per year, significantly exceeding the costs of designating and managing the network, then estimated at €5.8 billion per year. However, it was also estimated that EU co-funding of Natura 2000 was only meeting 9–19% of the financing needs and that national co-funding was not filling the remaining gaps.

The evaluation concluded that the Directives are not a barrier to sustainable development consistent with the conservation objectives of the sites. In the vast majority of cases, project proposals subject to Article 6 of the Habitats Directive permitting procedures are authorised. It did, however, identify that permitting delays can result from limited access to and availability of data and information. This can complicate the process of carrying out appropriate assessments and result in unnecessary burdens and costs to project developers. Sometimes those responsible for implementing the Directives, especially at regional and local levels, were not sufficiently aware of the requirements of or the flexibility provided by the Directives.

However, the evaluation revealed that there was a growing body of smart implementation approaches, such as early screening, better access to data and joined-up procedures between different permitting requirements, that can help reconcile environmental objectives with socio-economic development while keeping costs and administrative burden down.

Relevance

The evidence showed that the Directives' objectives remain fully valid and that they are still highly relevant to tackling the pressures and threats that species and habitats of EU conservation concern are facing. They set out what needs to be achieved and left it to the Member States to identify and respond as required to specific threats.

There were diverging opinions about whether the Annexes of species protected under the Directives were sufficiently updated, with some stakeholders questioning the listing of a small number of strictly protected species that have extended their range or that may be sufficiently abundant in a Member State to justify a lower level of protection. Other stakeholders argued that conflicts between the protection of these species and human activities could be fully addressed, where justified, using the flexibility provided by derogations under the Directives.

The evaluation concluded that the Annexes in their current form provide an adequate level of protection to enable the objectives of the Directives to be achieved. It also highlighted the complexity of amending the Annexes and noted that changes to the trigger species and habitat types for the selection of Natura 2000 sites could introduce significant legal uncertainty at a crucial time when establishment of the network was being finalised.

Coherence

Whereas the evaluation concluded that the Directives are coherent with each other it was recognised that there is a need to promote implementation solutions that optimise attainment of their conservation objectives while having full regard to the socioeconomic context in which they operate. This required strengthened dialogue and reinforced cooperation with the people who are involved with or impacted by the implementation of the Directives.

Although there are differences in specific objectives and scope of application between the Nature Directives and other related EU environmental directives (e.g. Environmental Impact Assessment and Strategic Environmental Assessment Directives (respectively, Directive 2011/92/EU, as amended, and Directive 2001/42/EC)) they were also considered consistent with and complementary to each other.

However, other sectoral EU policies have sometimes competing objectives. Given the significance of agricultural policy and the continuing decline of species and habitats associated with agriculture, an important conclusion was that greater efforts are needed to conserve and enhance biodiversity through more effective integration with the Common Agricultural Policy (CAP).

EU added value

The evaluation revealed that EU-level action through the Nature Directives remains fully valid and justified. The legal framework has created a more consistent, fair and integrated approach to nature conservation and delivery of ecosystem services across the EU.

Detailed implementation is regulated under national and regional laws and decisions in Member States according to their governance systems, supported by guidance and other EU-level policy documents.

The EU Action Plan for nature, people and the economy

The conclusions of the Nature Fitness Check provided a very strong impetus for action to enhance implementation of the Nature Directives. In its orientation debate on 7 December 2016, the College of Commissioners agreed to deliver an Action Plan as follow-up, in partnership with Member States, local authorities and key stakeholder groups. This would set out actions for enhanced and smarter implementation, including through guidance and regional outreach, promoting structured dialogue through stakeholder platforms, making best use of new technologies, and incentivising national and regional investment.

The action plan was prepared by a project team of 10 Commissioners, with close involvement of the European Committee of the Regions, in recognition that many implementation challenges and solutions require the active engagement of regional and local authorities. It was published in April 2017,[25] setting out an ambitious work programme covering four priority areas of action, identified as essential for better implementation of the Nature Directives and opportunities for better engagement of stakeholders and the public with the aim of building bridges between nature, people and the economy. It involved 15 main actions and over 100 individual measures to be achieved by the end of 2019. Whereas most measures were delivered within this timeframe, a small number of actions were completed afterwards.

Priority A. Improve guidance & knowledge and ensure better coherence with broader socio-economic objectives

This responded to the findings that implementing approaches, including inflexible application of the species protection rules, delays and unduly heavy burdens in site permitting procedures and insufficient stakeholder awareness can create needless tensions between nature protection and socioeconomic activities. The action plan promoted practical solutions, including smarter participatory approaches aimed at better involvement of landowners and users. It entailed a set of actions and measures to:

- Update, further develop and actively promote in all EU languages guidance on site permitting procedures, species protection and management, sectoral

guidance (e.g. on wind and other renewable energies and Natura 2000) as well as guidance on integrating ecosystem services into decision making;
- Fill knowledge gaps and promote enhanced quality and access to data;
- Establish a peer-to-peer support mechanism to help Member States' authorities on the site and species permitting requirements of the Directives.

There has been particular focus on improving and promoting guidance to support better implementation of the Nature Directives. By the end of 2020 the Commission had published updated guidance on "Managing Natura 2000 sites: The provisions of Article 6 of the 'Habitats' Directive 92/43/EEC"[26] and on wind energy developments and EU nature legislation[27] as well as new guidance on the requirements for hydropower in relation to EU nature legislation[28] and on energy transmission infrastructure and EU nature legislation. Updated methodological guidance documents on Articles 6(3) and 6(4) of the Habitats Directive and on species protection under the Habitats Directive were also at an advanced stage of preparation, and have since been published.[29]

Priority B. Building political ownership and strengthening compliance

This aimed to significantly boost implementation of the Directives, particularly with a view to ensuring that a fully coherent and functional Natura 2000 network is in place and that species are protected and, where appropriate, used sustainably. It involved the Commission working closely with Member States' competent authorities, who are ultimately responsible for implementing the Directives. This must be seen in conjunction with other measures in the broader area of environmental policy, such as the training of national judges and prosecutors, access to justice, and assuring compliance with EU environmental law. For example, the 2018 Commission Action Plan on environmental compliance and governance[30] directly cross-links to the Nature Action Plan. Priority B of the latter entailed a set of actions and measures to:

- Complete the establishment and ensure the effective management of the Natura 2000 network, including through development and delivery of conservation objectives and measures, ideally using management plans;
- Within the framework of the Environmental Implementation Review process to set up a structured bilateral dialogue process with Member State competent authorities to strengthen implementation of the Nature Directives and to discuss implementation challenges with landowners and other key stakeholders;
- Strengthen cooperation and share experience and expertise between Member States and stakeholders on the management of Natura 2000 at Biogeographical level;
- Further develop species action plans for the most threatened species and extend this approach to habitat action plans.

The introduction of nature dialogues has provided an important new framework to discuss with individual Member States the structural obstacles that impede achievement of the objectives of the Directives and to identify actions to overcome them. The dialogues have also provided a forum for the private sector, NGOs and other stakeholders to discuss implementation challenges and solutions. Implementation roadmaps have been agreed with the competent authorities in each Member State that have engaged with the process. By the end of 2020 nature dialogues had taken place in 24 Member States, with only Austria, Ireland and Poland remaining to have such discussions. A nature dialogue subsequently took place with Ireland in March 2021.

Priority C. Strengthening investment in Natura 2000 and improving synergies with EU funding instruments

This responded to the key finding that insufficient availability and uptake of funds are seriously impeding implementation of the Directives. It aimed at promoting better use of the available EU funding but also at examining ways of making nature more attractive for private investment. It also set out to prepare strategically for better uptake of nature funding requirements in different EU funds under the 2021–2027 multiannual financial framework. This involved a set of actions and measures to:

- Launch a process with Member States to update and improve the quality of Prioritised Action Frameworks (PAFs) to strategically identify funding needs and priorities for investment in Natura 2000 and related Green Infrastructure[31];
- Increase the Nature and Biodiversity envelope under the LIFE budget by 10% (probably equivalent to an additional €100 million for nature projects)[32];
- Stimulate private sector investment in nature through developing pilot projects to test the applicability of tools for private land conservation in Natura 2000;
- Increase awareness of opportunities and promote synergies with the major EU sectoral funds (CAP, Cohesion and European Maritime and Fisheries Funds);
- Provide guidance to support deployment of EU-level Green Infrastructure for better connectivity of Natura 2000 areas.

Given that the PAFs are strategic planning tools for strengthening investment in Natura 2000, particularly through use of EU funding instruments, a key focus was on developing an improved PAF format to better define needs and priorities. The Nature Directives Expert Group (NADEG) approved this in 2018. By the end of 2020 updated PAFs, some of which were still in a draft form, had been received from the majority of Member States. This has led to a significantly revised estimate of EU-wide financing needs for implementing the EU nature directives and related Green Infrastructure, calculated to be in the order of 15 billion €/ year for EU27. Training seminars were also organised in 22 Member States to discuss the possibilities for financing Natura 2000 in the multiannual financial framework 2021–2027 with the authorities and interested stakeholders.

Priority D. Better communication & outreach, engaging citizens, stakeholders & communities

This responded to the need to promote involvement and exchange of knowledge for the public, stakeholders, local authorities and communities as well as to give opportunities to youth to become more involved in the implementation of the Directives, using existing platforms. As most EU citizens are still not aware of Natura 2000, it also aimed to increase awareness of the importance of the network and provide greater recognition of good management practices in Natura 2000 areas. This entailed actions and measures to:

- Support recognition of good management of Natura 2000, including further development of the Natura 2000 award and establishing a Natura 2000 Day, as well as examining the potential of systems aimed at recognising positive management of sites
- Strengthen the links between natural and cultural heritage, particularly in the framework of 2018 as European Year of Cultural Heritage;
- Support knowledge exchange and engagement of local and regional authorities through a joint platform with the Committee of the Regions;
- Provide opportunities to youth under the European Solidarity Corps to develop their skills and experience and contribute to the conservation of Natura 2000 sites.

Key milestones in promoting awareness about Natura 2000 were the institution of a 'Natura 2000 Day' and further development of the Natura 2000 biannual award, which recognises the outstanding achievements of individuals and organisations that manage Natura 2000. Strengthening the links between Natura 2000 and cultural heritage[33] involved publication of case studies as well as a report on "Natural and cultural heritage in Europe: Working together within the Natura 2000 network" which includes a review of spatial overlaps between Natura 2000 and UNESCO World Heritage sites. A scoping study on tourism and recreational activities in Natura 2000 was also prepared, with a view to developing guidance on this subject.

The role of the Nature Directives and Natura 2000 in post-2020 EU biodiversity policy

The Nature Fitness Check evaluation confirmed the continued high relevance and importance of the Habitats and Birds Directives, especially of the Natura

2000 network, in EU biodiversity policy and the follow-up Action Plan for Nature, People and the Economy has given a real boost to their implementation.

Despite this progress, there is a need for better implementation to achieve the objectives of the Nature Directives. This is clearly underlined by the 2020 'State of Nature in the EU' assessment, which reveals the continuing decline of many protected species and habitat types of EU conservation concern. The assessment also showed some improvements, where targeted conservation actions have been taken at a sufficient scale.[34] A 2019 study for the Commission,[35] based on a detailed examination of genuine improvements recorded in the 2015 'State of Nature in the EU' assessment, offers valuable insights into the key drivers of success for Natura 2000, which are:

- Political support, strong and coherent governance and effective supporting institutions with motivated staff;
- Adequate and effective consultation and engagement of landowners and stakeholders;
- Research and monitoring that provides context-relevant information on the ecological requirements of targeted species and habitats and the pressures affecting them;
- Access to funding and other resources with the LIFE fund being a major catalyst for improvements, particularly when supported by follow-up investments under larger scale funds such as through EU agri-environment schemes;
- Ensuring long-term sustainability of conservation outcomes through commitments that provide on-going delivery of necessary measures.

There is also greater awareness now about the importance of biodiversity and the increased evidence of its decline, unprecedented in human history. The extreme rate of loss of biodiversity has been described by scientists as a 'sixth mass extinction.'[36] This is most compellingly illustrated by the 2019 global assessment of the Intergovernmental Scientific-Policy Platform on Biodiversity and Ecosystem Services (IPBES),[37] which confirmed that the accelerating rate of species extinctions and deteriorating health of ecosystems around the world, eroding food security, health and quality of life, represents an ecological crisis that has serious implications for society and our economies.

In her political guidelines for the European Commission, 2019–2024,[38] Commission President Ursula von der Leyen, included a 'European Green Deal' as one of her six headline ambitions for the European Union over the next five years and beyond. This recognises that we need to change the way we produce, consume and trade and that preserving and restoring ecosystems needs to guide all the work of the Commission. The political guidelines stated that we must set new standards for biodiversity cutting across trade, industry, agriculture and economic policy.

The European Green Deal, adopted by the Commission on 11 December 2019, responds to this challenge and sets out a new growth strategy for the EU.

It aims to transform the EU into a modern, resource-efficient and competitive economy where there are no net emissions of greenhouse gases by 2050, where economic growth is decoupled from resource use and where no one and no place is left behind. Among its strategic objectives is the protection, conservation and enhancement of the EU's natural capital. To achieve this goal the Commission undertook to present a new Biodiversity Strategy for 2030[39] that would include commitments, underpinned by measurable objectives, to address the main causes of biodiversity loss. It would also set out the EU's ambition for the 15th Conference of the Parties to the Convention on Biological Diversity, which concluded in Montreal, Canada, in December 2022 and at which the global biodiversity targets for 2030 were agreed.

The European Commission adopted the new EU Biodiversity Strategy, together with an associated action plan, in May 2020. This sets out a comprehensive, ambitious, long-term plan for protecting nature and reversing the degradation of ecosystems in the EU. The Strategy, which forms an integrated part of a broader action agenda under the Green Deal, including the EU Farm to Fork Strategy[40] that was adopted on the same day, sets out over 100 concrete actions under four areas of action, to tackle the loss of biodiversity. These are protecting nature, restoring nature, enabling transformative change, and EU action for an ambitious global agenda. The two areas of action most directly linked to implementation of the Nature Directives relate to expanding the network of protected areas and to the restoration of nature in the EU.

Providing a sufficient quantity and quality of space for nature is critical to the long-term viability of populations of different species and the health of ecosystems, particularly in the face of climate change.[41] The strategy recognises the essential role of protected areas and that the current network of legally protected sites, including those under strict protection, is not sufficient to safeguard biodiversity. It therefore sets out to expand the protected area network in the EU to cover at least 30% of the land and 30% of the sea, building on the Natura 2000 network, but also embracing national protected areas and integrating ecological corridors to achieve a coherent network. Having regard to the ecological requirements of certain habitats and species in areas of high biodiversity value, at least one-third of the protected area network, including areas important for mitigation and adaptation to climate change, and all primary and old growth forests, will need to be placed under strict protection.

It will be at least as important in the coming period to ensure that the protected areas are not just paper parks. There will be a need for outcome-based targets to ensure that the biological values of sites (species, habitats and ecological processes) are documented, managed and restored to ensure they achieve their conservation goals.[42] The strategy underlines the importance of effective management of the sites, based on clearly defined conservation objectives and measures and subject to appropriate monitoring. According to the strategy, the Commission, in collaboration with Member States and the European Environment Agency, was to provide criteria and guidance for identifying and designating additional protected areas, including a definition of strict protection, as well as for appropriate management planning. In doing so, it would explain how other effective area-based conservation measures and greening of cities can contribute to this objective. This formed the basis for a new Commission Staff Working Document – '*Criteria and guidance for protected areas designations*' – published in January 2022.[43]

The Biodiversity Strategy also aims to deliver an ambitious new EU Nature Restoration Plan for the coming decade. In this framework, the Commission undertook to put forward a proposal for legally binding EU nature restoration targets aimed at restoring degraded ecosystems, with particular focus on those with the greatest potential to capture and store carbon and to prevent and reduce the impact of natural disasters.[44] The Restoration Plan of the Biodiversity Strategy also aims to strengthen implementation of the Nature Directive by the Member States within clear deadlines. The target is to halt further deterioration in the conservation status of protected species and habitat types and to ensure that at least 30% of species and habitats not currently in favourable status are in that category or show a strong positive trend by 2030. The Commission recently provided guidance to Member States on how to select and prioritise species and habitats for this purpose.[45]

The EU and its Member States are well placed to make very significant progress on the restoration of species and habitats protected under the Nature Directives and in making Natura 2000 a fully functional network that delivers multiple benefits for nature, climate, people and the economy in the post-2020 period.

First, there is a current focus of the Commission and the Member States, aided where necessary through legal enforcement action, on identifying and establishing conservation objectives and the necessary measures for all the Natura 2000 sites. This is providing a greater level of clarity as to precisely what types of management and restoration actions are needed to achieve the favourable condition of the sites. In the EU Biodiversity Strategy, the Commission has committed to stepping up implementation and enforcement of EU environmental legislation. In relation to the Nature Directives enforcement will focus on finalising establishment of the Natura 2000 network, effective management of the sites, the species-protection provisions, and on species and habitats that show declining trends.

Second, the 2020 'State of Nature in the EU' assessment for the 2013–2018 period provides an essential baseline on the status and trends of protected species as well as an assessment of the contribution of Natura 2000 to achieving favourable conservation status. This will allow for the setting of credible nature restoration targets. The latest assessment conservatively estimates that 215,000 km^2 of habitats protected under the Habitats Directive need to be restored or actively recreated, both within and beyond Natura 2000 sites. This applies in particular to forests, grasslands, bogs, mires and fens, as well as coastal habitats, all of which have a vital role to play in mitigating the effects of climate change. In this context, the management and restoration of Natura 2000 have a particularly important contribution to make,[46] although its role and that of wider nature-based solutions have not yet been fully realised in climate policy.

Thirdly, the updating by Member States of their Prioritised Action Frameworks (PAFs) provides a strategic planning tool to help strengthen future investment in nature. The new format established for the PAFs will allow for more quantitative information on the species and habitats to be subject to active management and restoration in the next funding period. This will assist Member States in setting quantified targets for the management and restoration of areas for species and habitats, especially in Natura 2000 combined with wider Green Infrastructure and by way of other species conservation approaches, including through use of different EU financial instruments.

Member States are therefore now much better placed to realise the targets set out in the EU Biodiversity Strategy for the recovery of nature in the EU on land and in the marine environment. Delivery of the targets will ultimately depend on securing adequate financial and human resources and successful engagement of different stakeholder groups. The new LIFE Regulation[47] has a significantly enhanced budget to support nature and biodiversity projects during the next multiannual financial framework.[48] This will include strategic nature projects (SNaPs) to help mainstream Natura 2000 and wider biodiversity objectives into other policies and funding programmes. However, LIFE funding will only provide a small proportion of the overall investment needs for biodiversity and therefore, securing the necessary funding for Natura 2000, wider green infrastructure and nature restoration in the EU's key sectoral funds, especially under the Common Agricultural Policy, will be essential for success in the post 2020 period.

Notes

1 The author is a retired European Commission official, having worked on nature and biodiversity policy in DG Environment between 1992 and 2021. He co-ordinated the fitness check evaluation of the Nature Directives and was also actively involved in development of the Action Plan for Nature People and the Economy. Many colleagues in the Commission have also been deeply involved in these initiatives, particularly Nicola Notaro, Stefan Leiner, Luisa Samarelli, Natalie Pauwels, Ian Jardine (on secondment from Scottish Natural Heritage) and Ute Goerres of the Nature Unit, under the direction of Humberto Delgado, Director of Natural Capital in DG Environment. The information and views set out in this article are those of the author and do not necessarily reflect the official opinion of the European Commission.

2 Stanley Cramp (1977) *Bird conservation in Europe*. Her Majesty's Stationery Office, London.

3 European Commission, 'Wise Use and Conservation of Wetlands' (Communication) COM(1995) 189 final, 29.05.1995.

4 European Commission, EU Biodiversity Strategy for 2030 'Bringing nature back into our lives' (Communication) COM (2020) 380 final.

5 The concept of favourable conservation status is set out in Article 1 of the Habitats Directive. For species it means that population dynamics and natural range are favourable and there is, and will probably continue to be, a sufficient area of habitat

to maintain populations. For habitat types it means that the natural range and areas it covers are stable or increasing, the necessary structure and functions exist and are likely to continue into the future and that the conservation status of its typical species is favourable.

6 Case C-435/92 *Association pour la Protection des Animaux Sauvages and others v Préfet de Maine-et-Loire and Préfet de Loire-Atlantique* ECLI:EU:C:1994:10.

7 Proposal for a Council Directive amending Directive 79/409/EEC on the conservation of wild birds. COM (94) 39 final – 94/0061 (SYN). OJ C 100, 9.4.1994, pp. 12–13.

8 European Commission, Guidance document on hunting under Council Directive 79/409/EEC on the conservation of wild birds (2008) http://ec.europa.eu/environment/nature/conservation/wildbirds/action_plans/guidance_en.htm.

9 Agreement between BirdLife International and FACE on Directive 79/409/EEC, 12 October 2004, https://ec.europa.eu/environment/nature/conservation/wildbirds/hunting/docs/agreement_en.pdf

10 C-3/96 Commission v the Netherlands ECLI:EU:C:1998:238, C-240/00 *Commission v Finland* ECLI:EU:C:2003:126, C-202/01 *Commission v France* ECLI:EU:C:2002:713, C-378/01 *Commission v Italy* ECLI:EU:C:2003:176, C-235/04 *Commission v Spain* ECLI:EU:C:2007:386, C-334/04 *Commission v Greece* ECLI:EU:C:2007:628, C-418/04 *Commission v Ireland* ECLI:EU:C:2007:780.

11 C-67/99 *Commission v Ireland* ECLI:EU:C:2001:432, C-71/99 *Commission v Germany* ECLI:EU:C:2001:433, C-220/99 *Commission v France* ECLI:EU:C:2001:434.

12 Marc Dufrêne and Michel Fautsch, *Natura 2000, une opportunité pour la nature en Wallonie* (Weyrich, 2008), p. 33.

13 Based on data provided by the Member States the European Environment Agency regularly updates a Natura 2000 barometer https://www.eea.europa.eu/data-and-maps/dashboards/natura-2000-barometer.

14 At the end of 2019 there were 934 Natura 2000 sites in the United Kingdom, covering a total area of 153,137 km^2, with 86% of the area relating to marine Natura 2000 sites.

15 Andrew Jackson, *Conserving Europe's Wildlife. Law and Policy of the Natura 2000 Network of Protected Areas* (Routledge 2018) 29.

16 European Commission, 'The State of Nature in the European Union Report on the status of and trends for habitat types and species covered by the Birds and Habitats Directives for the 2007–2012 period as required under Article 17 of the Habitats Directive and Article 12 of the Birds Directive' (Report) COM/2015/0219 final. This report was underpinned by a detailed technical assessment of the European Environment Agency (EEA). Technical report No 2/2015 https://www.eea.europa.eu/publications/state-of-nature-in-the-eu.

17 European Commission, 'Regulatory Fitness and Performance (REFIT): Results and Next Steps' (Communication) COM(2013) 685 final.

18 European Commission, 'Fitness Check Mandate for Nature Legislation', 25 February 2014, https://ec.europa.eu/environment/nature/legislation/fitness_check/docs/Mandate%20for%20Nature%20Legislation.pdf

19 Milieu, IEEP and ICF, Evaluation Study to support the Fitness Check of the Birds and Habitats Directives, March 2016. http://ec.europa.eu/environment/nature/legislation/fitness_check/docs/study_evaluation_support_fitness_check_nature_directives.pdf

20 European Commission, 'List of Key Documents' (undated), http://ec.europa.eu/environment/nature/legislation/fitness_check/docs/List%20Key%20documents.pdf

21 Elena Fries-Tersch, Kerstin Sundseth and Marta Ballesteros, 'Report on the Open Public Consultation of the 'Fitness Check' on the Birds and Habitats Directives', Final report for the European Commission, DG Environment. Brussels, October 2015. http://ec.europa.eu/environment/nature/legislation/fitness_check/docs/consultation/public%20consultation_FINAL.pdf

22 European Commission, 'Fitness Check of the EU Nature Legislation (Birds and Habitats Directives)' (Staff Working Document) SWD (2016) 472 final.
23 European Commission, 'Investing in Natura 2000: Delivering Benefits for Nature and People' (Staff Working Paper) SEC(2011) 1573 final.
24 European Court of Auditors, Special Report No 1/2017: More efforts needed to implement the Natura 2000 network to its full potential. https://www.eca.europa.eu/Lists/ECADocuments/SR17_1/SR_NATURA_2000_EN.pdf
25 European Commission, 'An Action Plan for nature, people and the economy' {SWD(2017) 139 final}' (Communication) COM(2017) 198 final.
26 European Commission, (Commission notice) C(2018) 7621 final.
27 European Commission, (Commission notice) C(2020) 7730 final.
28 European Commission, (Commission notice) C(2018) 2619.
29 European Commission, 'Assessment of plans and projects in relation to Natura 2000 sites – Methodological guidance on Article 6(3) and (4) of the Habitats Directive 92/43/EEC' C(2021) 6913 final, and European Commission, 'Guidance document on the strict protection of animal species of Community interest under the Habitats Directive' (Commission Notice) C(2021) 7301 final.
30 European Commission, 'EU actions to improve environmental compliance and governance {SWD(2018) 10 final}' (Communication) COM(2018) 10 final.
31 Green infrastructure is a strategically planned network of natural and semi-natural areas with other environmental features designed and managed to deliver a wide range of ecosystem services such as water purification, air quality, space for recreation and climate mitigation and adaptation. The Natura 2000 network constitutes the backbone of the EU green infrastructure.
32 This was achieved through Commission Delegated Regulation (EU) 2018/93 of 16 November 2017 on the increase of the percentage of the budgetary resources allocated to projects supported by way of action grants under the sub-programme for Environment dedicated to projects supporting the conservation of nature and biodiversity according to Article 9(4) of Regulation (EU) No 1293/2013 of the European Parliament and of the Council on the establishment of a Programme for the Environment and Climate Action (LIFE) and repealing Regulation (EC) No 614/2007
33 https://ec.europa.eu/environment/nature/natura2000/management/links_natural_cultural_heritage_en.htm
34 European Commission, The state of nature in the European Union. Report on the status and trends in 2013–2018 of species and habitat types protected by the Birds and Habitats Directives, (Report) COM/2020/635 final.
35 Graham Tucker, Tom Stuart, Sandra Naumann et al, 'Study on Identifying the Drivers of Successful Implementation of the Birds and Habitats Directives', Report to the European Commission, DG Environment on Contract ENV.F.1/FRA/2014/0063, Institute for European Environmental Policy, Brussels (2019).
36 For example see Gerardo Ceballosa, Paul R. Ehrlich and Rodolfo Dirzob, 'Biological Annihilation Via the Ongoing Sixth Mass Extinction Signaled by Vertebrate Population Losses and Declines' (2017) 114(30) *Proceedings of the National Academy of Sciences* E6089–E6096; DOI: 10.1073/pnas.1704949114.
37 IPBES, 'Summary for Policymakers of the Global Assessment Report on Biodiversity and Ecosystem Services of the Intergovernmental Science-Policy Platform on Biodiversity and Ecosystem Services' (2019) https://doi.org/10.5281/zenodo.3553579.
38 Ursula von der Leyen, 'A Union that Strives for More: My Agenda for Europe. Political Guidelines for the Next European Commission' (2019) https://www.europarl.europa.eu/resources/library/media/20190716RES57231/20190716RES57231.pdf
39 European Commission, The European Green Deal (Communication) COM/2019/640 final
40 European Commission, A Farm to Fork Strategy for a fair, healthy and environmentally-friendly food system (Communication) COM/2020/381 final

41 The famous Harvard Biologist, E.O. Wilson argued for setting aside half of the Earth for nature conservation in order to manage sufficient areas to prevent species extinctions and ensure the long-term health of ecosystems (*Half-Earth: Our planet's fight for life* (Liveright 2016)). Such a level of ambition has generated controversy and there are divergent views on the need for, desirability and feasibility of achieving it (Bram Buscher and Robert Fletcher, 'Why E O Wilson Is Wrong About How to Save the Earth' *Aeon*, 1 March 2016, https://aeon.co/ideas/why-e-o-wilson-is-wrong-about-how-to-save-the-earth). Some biologists argue that the key issue for biodiversity is to prioritise the right places rather than just the total area protected (Stuart L. Pimm et al, 'How to Protect Half of Earth to Ensure it Protects Sufficient Biodiversity' (2018) 4(8) *Science Advances* (29 August 2018)). However, Christina Pasca Palmer, the then Executive Secretary of the Convention on Biological Diversity proposed an expansion of nature reserves, ocean protected areas, restoration projects and sustainable land use regions by 10% every decade (Jonathan Watts, 'Make Half of World More Nature-Friendly by 2050, Urges UN Biodiversity Chief' *The Guardian* (London, 13 April 2018), https://www.theguardian.com/environment/2018/apr/13/make-half-the-world-more-nature-friendly-by-2050-says-un-chief)

42 This will need to be supported by appropriate indicators to measure progress (e.g. species abundance, area and condition of habitat) as highlighted by Piero Visconti et al, 'Protected Area Targets Post-2020' (2019) 364(6437) *Science* (19 April 2019) 239–241.

43 European Commission, 'Criteria and Guidance for Protected Area Designations' (Staff Working Document) 28.1.2022, SWD(2022) 23 final.

44 European Commission, 'Proposal for a Regulation of the European Parliament and of the Council on nature restoration', 22.6.2022, COM(2022) 304 final.

45 European Commission, 'Biodiversity Strategy for 2030: Guidance to Member States on how to Select and Prioritise Species/Habitats for the 30% Conservation Improvement Target Under the Strategy' (2021) https://circabc.europa.eu/ui/group/fcb355ee-7434-4448-a53d-5dc5d1dac678/library/4d8f2f91–7708-4ed2-ba0e-e7a945a6d56a/details.

46 European Commission, 'Guidelines on Climate Change and Natura 2000. Dealing with the Impact of Climate Change on the Management of the Natura 2000 Network of Areas of High Biodiversity Value' Technical report – 2013 – 068 (2013).

47 European Commission, Regulation (EU) 2021/783 of the European Parliament and of the Council of 29 April 2021 establishing a Programme for the Environment and Climate Action (LIFE), and repealing Regulation (EU) No 1293/2013, OJ L172.

48 European Commission, 'EU Budget: Commission Proposes to Increase Funding to Support the Environment and Climate Action' (Press release), 1 June 2018, http://europa.eu/rapid/press-release_IP-18–4002_en.htm

3

THE REQUIREMENTS FOR AN EFFECTIVE INTERNATIONAL NATURE CONSERVATION SYSTEM, AND THE EU BIRDS AND HABITATS DIRECTIVES

John Temple Lang

An international nature conservation policy, to be effective, must as far as possible fulfil a number of requirements.[1] Although they are interconnected, it is useful to consider them separately.

Important areas

Everywhere in the world, but in particular in densely populated regions such as Europe, it is necessary to identify the areas of international importance for the species to be protected and to protect, and if necessary to restore, the habitats in those areas. These areas may be important for breeding, wintering, feeding, roosting, or moulting species, or may simply be areas where the species gather or are obliged to gather in bottlenecks while on migration.

These areas must be identified on the basis of objective scientific criteria, which necessitate quantitative and qualitative information about the areas to be considered. This information must be kept up to date, as habitats alter naturally even when they are not altered by human activities. An area that was of limited importance for the species in question may become more important if other similar areas are damaged and the populations that previously relied on them are obliged to move to the area that was previously less important. Research on migratory species using a given area may show, for example, that although the numbers of the species involved at any one moment are limited, there is a high turnover of individuals using an area, so that the total number of individuals depending on it may be much larger than appears at first. Research may also show that an area may be of particular importance because it is where migrants need to obtain large amounts of food to enable them to complete the next stage in their journey, which may involve, in the case of many bird species in different parts of the world, continuous flights of thousands of kilometres, with no possibility of resting or feeding on the way.

DOI: 10.4324/9780429299100-3

Assuming that a sufficient amount of quantitative and qualitative information is available about the areas to be assessed, it is necessary to identify the most important areas. This may be done in one or more ways.[2] An area may be regarded as being of international importance because it is the only place or one of a very small number of places where a rare species or an unusual habitat occurs, even if the number of individuals found there is small, because of the need to give protection to the species or habitat in question. An area may be of special importance because it is the best or the best preserved example of a particular habitat. More generally, it is usual to regard as being "of international importance" areas that regularly hold more than a given percentage, traditionally 1%, of the biogeographical population of the species to be protected. That population may be less than the world population of the species but may need to be conserved if it is separate from populations elsewhere. This criterion assumes that it is possible for scientists to estimate the whole of the biogeographical population part or all of which occurs in the area. Other, non-quantitative, criteria and priorities can be based on vulnerability, rarity and naturalness.[3]

Conservation of areas of international importance

When the areas fulfilling the selection criteria have been identified, the areas must be safeguarded and managed in such a way as to maintain and if necessary to restore their value to the species to be protected. As a minimum, this almost always means protecting them from industrialisation or other physical development or change, from pollution, and from serious disturbance.[4] The conservation measures that are needed will depend on the habitats or species to be protected, the nature and seriousness of whatever threats there may be, and the willingness or otherwise of the landowners (or in the case of marine areas, the fishermen or oil and gas companies or other users) to cooperate with whatever conservation measures are needed. This means, in short, that for every area identified there must be at least conservation objectives and the essential elements of a management plan, even if it consists only of a list of actions that must be avoided or prevented. Any such objectives and management plan must be written specifically for the area in question: there is no such thing as useful generic objectives or a useful all-purpose plan. Indeed, a recent study published in *Nature* re-emphasises the importance of well-managed protected areas, demonstrating that management for (in this case) waterbirds was consistently positively correlated with protected area success.[5]

Any such plan must deal with pollution threats, disturbance and deliberate or accidental killing of the species to be protected.

The need for a monitoring body

Experience shows that States cannot be relied on to do everything that they should do to conserve species or habitats in their territories. To be effective, an international nature conservation system needs a monitoring body that is separate from and independent of the participating States. The tasks of such a body are to ensure, as far as possible, that each State recognises the importance of the areas in its jurisdiction (including its territorial seas), ensures that management plans are adopted and carried out, and that any development that would damage any area is prevented or halted or altered so as not to cause damage. Such a body need not necessarily be an inter-governmental body: indeed, it may be more effective if it is not, because inter-governmental bodies often lack the independence necessary to insist on action being taken to conserve a threatened area. What is required is a degree of independence, and the resources sufficient to enable it to do its job.

Enforcement

Traditional international agreements provide for enforcement by negotiation and, if necessary, by withholding the benefits of the agreement from a State that is not fulfilling its obligations. Threats of withholding benefits are ineffective in the case of nature conservation agreements unless the agreement concerns the shared exploitation of a resource such as a fishery. So enforcement necessitates regular meetings of the participating States, with agendas that the monitoring body can ensure raise all the issues that need to be dealt with. Regular meetings, the discussions of which are publicly known, provide valuable opportunities for non-governmental organisations, the public at large, and the monitoring body to put pressure on any States that are not fulfilling their obligations. In particular, it is important that regular meetings are designed to ensure that economic pressures to weaken or abandon habitat conservation measures are resisted.

Another kind of enforcement mechanism can be provided by setting up an advisory body to consider complaints and to say whether the States' obligations have been complied with, fully or at all, and to make recommendations regarding how compliance may be achieved. Provided that such a body is sufficiently independent and adequately resourced, this can be valuable, as the Compliance Committee under the Aarhus Convention has shown.

The effectiveness of nature conservation policies does not depend only on the enforcement of nature conservation legislation: it also depends on agricultural,[6] fisheries, forestry and land use policies, and on measures to control or prevent pollution. An effective nature conservation policy cannot be based only on specially protected areas, and so all of the policies that influence the environment outside as well as inside those areas will determine whether the policy is effective or not. The ecological consequences of policies adopted for economic purposes are often ignored.

In addition, all nature conservation policies require effective enforcement of laws regulating or prohibiting hunting or fishing, to protect vulnerable species or to prevent over-exploitation, and to prevent interference with reproduction. It may be necessary, for example, to prohibit hunting at or before the breeding season, when populations are at their lowest and the impact of hunting on breeding numbers is at its greatest.

A court or courts

Ideally, the obligations of States under an international nature conservation agreement should be interpreted and enforced, when necessary, by a court. Such a court would not need to be specially set up for the purposes of the agreement: an international court, or the court of a regional economic integration organisation, might be sufficient.

There are at least two roles that such a court could have, depending on its structure and relations with the participating States. First, it could hear cases brought against States that are said not to be fulfilling their legal obligations under the agreement. Such cases could be brought by other States, although this is not certain to work effectively: States are reluctant to embarrass one another. Second, a court could decide cases brought by e.g. non-governmental organisations, against national authorities or private interests if their actions were contrary to the agreement. Such a court could be a national court, applying the international agreement directly as part of national law, or it could be an international court such as the Court of Justice of the European Union (CJEU), to which national courts refer questions of European Union law for authoritative rulings. The essential requirement is that non-governmental bodies should be able to have cases considered by a judicial authority.

The effectiveness of any international nature conservation system depends greatly, among other things, on the opportunities that it provides to non-governmental organisations to initiate legal proceedings, as distinct from merely lobbying and exercising political and other pressure through the media.

Resources, political will and scientific information

Whatever the other characteristics of an international nature conservation system may be, the success or failure of any such system will depend to a large extent on whether the participating States are willing to provide the resources in funds and personnel to make the system work. Various examples could be given, even in States with relatively good reputations for conserving nature, of reorganisations of national authorities and appointment of individuals the objectives of which were clearly to make nature conservation ineffective.

All nature conservation policies require scientific information, on the populations of the species that may need to be conserved, and on the ecology of the species and habitats, so that conservation measures can be carefully designed. The more scientific information is available, and the more research to obtain more information is ongoing, the more effective the policy is likely to be. Non-governmental organisations can be useful suppliers of scientific information, especially information that can be obtained only by large numbers of observers.

The European Union Birds Directive in the light of these requirements

It is now appropriate to assess the EU Birds Directive, which was adopted in 1979, in the light of these requirements. Since then it has been changed only in a limited number of respects.[7]

Almost from its inception, it was understood that the most important provisions of the directive were those dealing with "special protection areas" for migratory bird species. The directive included an important but general obligation to conserve habitats for birds outside specially protected areas, but this has not proved to be very successful.[8] The directive also deals with illegal killing and trapping of birds, but these provisions depend for their effectiveness on enforcement by national police, and this has not always been provided.

The key habitat provision of the directive was Article 4, which required EU Member States to establish a network of specially protected areas (SPAs) for the conservation of certain endangered and migratory bird species. This was the extension of the concept of "areas of international importance" in the Ramsar Convention on wetlands, as defined by the conference in Cagliari of the parties to the Ramsar Convention. The Birds Directive required the Member States to list the important areas in their jurisdictions, but the States initially listed only the important areas that they intended to protect, or that they already protected. A standard of comparison was needed, for assessing whether the Member States' lists were complete. BirdLife International, as it now is, prepared a series of lists of areas of international importance, using the Cagliari criterion of 1% of the biogeographical population of each species.[9] The novelty of this approach was its extension to a broader group of bird species (the Ramsar Convention applied only to waders and waterfowl) and its extension to areas other than breeding and wintering areas. Each area listed was described and the numbers of the species for which the area was important were given. The original list was extended to more States and repeatedly revised in the light of additional information.[10]

The result was that in a case brought by the European Commission against the Netherlands in the CJEU, the Court accepted that the BirdLife list was the best scientific information available, and that special protection should be provided in each of the areas listed in the Netherlands, unless there was evidence to the contrary.[11] This implied, of course, that all the other EU Member States should do the same in their jurisdictions.

The BirdLife list of important bird areas now consists of two volumes for Europe.[12] The relatively simple "one percent of the biogeographical population" has been supplemented by introducing priorities but has not been fundamentally altered.

More important, the idea of a list of "important bird areas" for the European Union has now been copied in lists of important areas for various taxa all over the world.

The role of the European Commission

One of the responsibilities of the European Commission is to ensure that EU Member States carry out their obligations under EU directives. Directives are legally binding measures that Member States are obliged to carry out by whatever national measures are appropriate. The States are legally obliged not only to enact the legislative and other measures that are needed to fulfil the obligations imposed by the directive but to enforce them and carry them out.

This responsibility has led the Commission to bring a number of cases in the CJEU against Member States. In addition, the Commission is in regular contact with the national authorities in all the Member States, with a view to ensuring that they are fulfilling their obligations.

The Commission is therefore, at EU level, the monitoring body for the Birds and Habitats Directives. In fact, it is the principal monitoring body for the whole sphere of European Union law, and therefore when it takes action to enforce these two directives, it is using the considerable powers that it exercises for many other purposes. The Commission therefore has the legal and political authority and weight, as well as the resources, to enforce the two directives. It also has a considerable degree of independence from the EU Member States.

In addition to its powers to deal directly with the national authorities, the Commission also has the responsibility of submitting observations to the CJEU on every question that is referred to the Court by national courts. This means that when cases come before national courts involving the interpretation of the Birds and Habitats Directives, or of the Environmental Impact Assessment Directive, and the national court wishes to get a ruling on any question of EU law, the Commission submits observations to the Court on the answer to be given. This is important because private parties, if they have standing to bring proceedings under national law, may rely on the EU directives in legal proceedings against national authorities in national courts.

The Commission's powers to bring States before the Court when necessary helps to ensure that the States take seriously whatever the Commission wants to discuss with them. Member States have legal obligations to give the Commission information about the application of the Birds Directive when asked to do so, and they are also required to report at intervals to the Commission. These obligations are not always or fully carried out, but they are important and necessary. Unlike, for example, the United States Fish and Wildlife Service, the Commission is not an EU conservation body, and is obliged to rely on the national authorities of EU Member States to put the Birds Directive into effect on the ground, both by ensuring conservation of habitats and by enforcing the rules about illegal killing.

As well as the European Commission, the Compliance Committee under the Aarhus Convention also plays a useful role, in spite of the fact that it has not got legal powers corresponding to those of the Commission. Indeed, the Aarhus Compliance Committee can be useful in particular when the Commission for one reason or another fails to take the steps needed to protect important areas, and when proceedings by non-governmental organisations in national courts may be insufficient. (The Aarhus Convention gives non-governmental organisations certain rights to bring proceedings in national courts to enforce environmental obligations.)

In short, the Commission has a key role in applying and enforcing the Birds Directive, but its role is carried out in the context of the institutions of the European Union. This makes it arguably the most important nature conservation body in the world.[13]

This also means that the Commission is almost automatically involved in international environmental discussions and negotiations outside the EU, and is in a position to influence them to an important extent.

The Commission has worked with the European Environment Agency as well as with Member State authorities and non-governmental organisations. Action plans for various species in special need of conservation have been produced. The most important way by which the Commission supplements its enforcement powers was the compilation and adoption of the Natura 2000 list of sites for protection under the Birds and Habitats Directives, and the provision of funds for acquisition and in particular management of Natura 2000 sites. There are now nearly 27,000 of these sites. Naturally, developed management plans have not yet been produced for all of these sites, but that is the next stage.

The importance of the list of important areas for birds under the Birds Directive

The BirdLife International list of the most important areas for birds in the EU is based on objective scientific criteria elaborately explained in the two volumes of the list.[14] It provides the basis for a network of nature reserves for endangered and migratory birds across the EU. It constitutes a comprehensive network of areas for breeding, wintering, feeding, moulting and roosting, and is intended to provide habitat protection at staging areas along migration routes. It is also intended to provide protection in places where birds are vulnerable to being shot at "bottlenecks" such as mountain passes and straits where migratory birds come together. This network has a purely scientific basis: it is not based on the national frontiers of States.

Ideally, it would be desirable to have similar lists for all migrating taxa – e.g. fish, cetaceans and insects. The recent development of a list of so-called Key Biodiversity Areas is a step in this direction.[15]

Enforcement of the Birds Directive

The Birds Directive is enforced largely through regular formal and informal contacts between the Commission and the national authorities of Member States. The Commission relies to a considerable extent on information provided by non-governmental organisations, as well as information provided (sometimes unwillingly) by the national authorities themselves. The fact that the Commission has funds in the Natura 2000 programme (via EU LIFE funding) to contribute to the cost of conservation is a necessary element in this cooperation. Nature conservation costs money, and the EU nature conservation measures would not work as satisfactorily if they were implemented only by legal enforcement. Enforcement against private parties is carried out by national authorities, and in several EU Member States it is clearly insufficient.

The Court of Justice of the European Union

The CJEU exercises two kinds of jurisdiction in connection with the Birds Directive, and the other EU directives concerned with nature conservation. First, the Court decides cases brought by the Commission against Member States for failure to carry out their obligations under the Directive, the Habitats Directive, or some other rules of EU law, such as the Environmental Impact Assessment Directive, the Water Framework Directive, or the Environmental Liability Directive. In such cases the Court has power to order interim measures against Member States (e.g. an injunction), although this power is not often used.

Second, as already mentioned, the Court answers questions asked by national courts in cases brought under EU law by private parties or non-governmental organisations, under national law and in accordance with the Aarhus Convention.[16] The Court's judgments in these cases are legally binding on the national courts in all the EU Member States. These judgments help to ensure that EU law is enforced at national level, and is interpreted and applied uniformly throughout the EU. The judgments of the Court are almost always followed and applied by national courts and authorities, albeit compliance may take some time.

The Court has been exercising both kinds of jurisdiction under EU law since it was set up in 1952, so it has established (if sometimes controversial) authority.

The Court has been enormously important in the working of the Birds Directive. The Court has interpreted and applied the Directive strictly and correctly, and has resisted a number of efforts by several Member States to weaken the Directive, or to allow economic considerations to prevail over the obligations to conserve nature. The Court has consistently provided the Commission with important legal support, and greatly strengthened the Commission's position in its dealings with Member States. Its rulings have also strengthened the position of non-governmental organisations in dealing with national authorities, and with companies whose activities were endangering nature conservation.

No other nature conservation body in the world has the benefit of judicial support of this kind. The Court has done a great deal to reduce one of the most serious weaknesses in the EU nature conservation system, its dependence on enforcement by national authorities.

For the Court, the Birds and Habitats Directives are merely two of the many EU directives that the Court is called on to interpret and apply. This means that the Court brings to the interpretation of these two directives the principles that it uses in other areas of law. One of the most important of these is the "precautionary principle", now stated in Article 191 TFEU, which in essence allows the EU to act "on the safe side" in situations where the scientific advice is unclear. The Court has said that

> where there is uncertainty as to the existence or extent of the risks to the health of consumers, the institutions may take protective measures without having to wait until the reality and the seriousness of those risks become fully apparent.[17]

In its Communication on this principle in 2000, the Commission said that to avoid risk it is justifiable to act

> where the scientific evidence is insufficient, inconclusive or uncertain and there are indications through preliminary objective scientific evaluation that there are reasonable grounds for concern that the potential dangerous effects on the environment, human, animal or plant health may be inconsistent with the chosen level of protection.[18]

This principle applies where the risks of harm to the environment are assessed under the Habitats Directive and/or the Environmental Impact Assessment Directive, and this illustrates the value of having environmental law as part of a broader legal system.

Another advantage of a broader legal system arose in the Slovak Brown Bear case.[19] The question before the Court was whether a non-governmental organisation could apply to a national court, even in the absence of any general directive on access to courts in environmental cases, to object to a breach of the Habitats Directive. The Court decided that there is a general principle of EU law that national procedural remedies for breaches of EU law must be effective. This principle is not limited to environmental law cases but is important in those cases.

The principle that national remedies must be effective results from another general principle of EU law with important results for nature conservation. This is the duty of "sincere cooperation", based on Article 4(3) TEU which obliges national authorities inter alia to take measures to ensure fulfilment of obligations arising from the EU Treaties or from acts of EU institutions.[20]

This principle has led the Court to rule that national courts must "give an interpretation of national procedural law which, to the fullest extent possible, is consistent with the objectives" of the Aarhus Convention on access to justice.[21]

Although the Court has not yet said so explicitly, it seems that the principle of sincere cooperation also means, in the light of the environmental provisions of the TEU, the TFEU and the Charter, that EU States must not allow a development that has been assessed as seriously harmful under the Environmental Impact Assessment Directive. If the Court were to confirm this, it would be extremely important.[22]

The European Union Habitats Directive in the light of these requirements

Since the EU Habitats Directive[23] is based on the same principles as the Birds Directive, it fulfils the requirements set out above to much the same extent and in substantially the same way. There are however a number of differences.

One of the most important differences concerns the criteria for selection of sites to be conserved. For the purposes of the Habitats Directive it was necessary to produce a classification of habitat types, and then to make lists of the most important examples of each type. Both operations involved expertise and judgement rather than the primarily quantitative criteria used for the purposes of the Birds Directive. The variety of habitats to be listed was very great, and much more information was available about some kinds of habitats than others. It was important to ensure, as far as possible, that some kinds of habitats were not neglected merely because they are less well known. Birds were much better known than most other organisms.[24]

Both the list of habitat types and the lists of sites to be protected had to be specially drawn up for the purposes of the Habitats Directive. There were no comprehensive Europe-wide lists on which the Directive could be based, even though much information was available in many and various published sources. The lists for the Directive were initially based on national lists, some of which were excellent,[25] as well as on the classification of European habitats developed by the CORINE biotopes project.[26]

The Habitats Directive obliged Member States to identify and designate "special areas of conservation" (SACs) on the basis that the sites represent one of the habitat types listed in the Directive, the size of the sites, the numbers of protected species in the site, and the extent to which the site can support those species. Although the Directive was intended to make the process of site selection objective and not discretionary, those aims were not (and probably could not have been) fully achieved. There was, therefore, much reliance on peer review by experts, and on comparisons between sites of the same habitat types in different EU States. Given the limitations of the inventories of important sites at the time when the Habitats Directive was agreed in 1991, it is not easy to see that the initial site selection process could have been much better planned (albeit several countries' early approaches to site designation were criticised for insufficient public consultation).[27] There was nothing available, for many of the habitat types listed in the Directive, corresponding to the large list of important areas for birds.

Writing from an international perspective, Griffiths and Vogiatzakis[28] said:

> There is no generally accepted scheme for establishing conservation priorities, partly as a result of the complexity and variety of life itself but also because of limited resources of time and money and the need to prioritize. The criteria established by Ratcliffe in the Nature Conservation Review in Britain...... have been developed and adopted by the International Union

> for the Conservation of Nature. However.... the information required to implement the criteria established by Ratcliffe often does not exist....

To put the Habitats Directive into effect, EU States identified sites that are important for habitats and species "of European interest" on purely ecological grounds, and according to agreed scientific criteria. The national lists were circulated and assessed, and States were asked to add more sites to complete the European list. The sites finally listed and adopted by way of formal Decisions of the European Commission should then be protected by national measures. This process was much less automatic than under the Birds Directive and much more difficult. A great deal of information was needed, much of it unavailable, and there were few comprehensive or authoritative lists available for any of the habitat types. Some States were better documented than others, and efforts were needed to bring all the States as far as possible to the same level.

In particular, not enough information was available about maritime habitats. Too little information was available about feeding areas for migratory birds (in the context of the Birds Directive) and cetaceans, and about benthic (sea-floor) habitats.

For the species covered by the Birds and Habitats Directives, the aim is to ensure that populations are maintaining themselves, their range is not shrinking, and there is enough habitat to maintain populations in the long term. For the habitat types, the aim is that the range and the sites within that range are at least stable, the sites provide what is needed to maintain or restore the habitat, and the conservation status of the species of Community interest in the habitat is favourable.

It is important that sites must be listed on the basis of their scientific and ecological importance: economic considerations must not influence the initial listing[29] even if those considerations may have to be taken into account later.

In the case of important areas under the Birds Directive, what was usually necessary was to avoid damage to the habitats involved. Conservation of important areas of habitat however usually necessitates policies or plans for the management of the areas, including not only safeguards against harm but also positive actions to maintain or restore ecological value. Harm, if it has occurred or is unavoidable, must be mitigated. Research was often needed to determine with confidence what was needed. In some cases acquisition or compensation was clearly the only long term solution, and funds were needed for that purpose also. It would appear that Member States sometimes felt it necessary to choose to conserve one area rather than another, not because it was more valuable, but because it was for example easier to protect it against economic development. Such an approach would, however, seem likely to infringe the requirement to propose sites under the Habitats Directive solely on the basis of the criteria set out in Annex III, such that the Commission has "an exhaustive list of the sites which, at national level, have an ecological interest which is relevant from the point of view of the Habitats Directive's objective of conservation of natural habitats and wild fauna and flora".[30] While the European Commission has won several cases against Member States regarding late or insufficient designation of sites under the Habitats Directive, establishing that a particular area should have been designated is more difficult under the Habitats Directive than under the Birds Directive because there is no authoritative list of sites in the context of the former, in contrast to the Birds Directive which has the Important Bird Area list, as discussed above.

Under the provisions of the Habitats Directive, industrial development and other economic considerations were given greater weight than under the strict requirements of the original terms of the Birds Directive, before its amendment by the Habitats Directive.[31]

Under the Habitats Directive, provided that no alternative is available,[32] and provided that favourable conservation status is maintained, Member States may depart from the normal protective regime for inter alia "imperative reasons of overriding public interest, including those of a social or economic nature". This is significantly weaker than the original terms of the Birds Directive, which the Court had interpreted strictly.[33]

The Commission has given national authorities some guidance about conservation objectives and measures that should be set and taken for each Natura 2000 site.[34] The objectives of the site (why it is important and for what species/habitat(s)) must be defined explicitly so that threats can be assessed and appropriate conservation measures developed. The need for sound information and assessment of the costs and benefits of possible conservation measures is stressed. This leaves a great deal to the judgement of national authorities, and to the thoroughness of assessments under the EIA and/or Habitats Directives. These are made only when a project is likely to have a significant effect on the environment (EIA) or when a plan or project is likely to have a significant effect on the Natura 2000 site in question (Habitats Directive), so such assessments are not a substitute for the management plans required by the Habitats Directive.[35] Further, as a matter of law, it would seem that an assessment under the Habitats Directive cannot be carried out lawfully in the absence of conservation objectives for the site since Article 6(3) requires that the assessment be made of "the implications for the site *in view of the site's conservation objectives*". The same is arguably true of management plans: that is, in light of the principle of sincere cooperation in Article 4(3) TEU, it is arguably necessary to develop or improve management plans in order to carry out an assessment under the Habitats Directive.[36]

Overall assessment of the EU nature conservation policy so far – some comments

Although this chapter here concentrates on the Birds and Habitats Directives, it must be remembered that both measures were being applied in Member States that were being altered by the EU's agricultural policy, fisheries policy and forestry policy. Among many other things, this meant that birds and other organisms that are not concentrated in particular areas were much less effectively protected, and indeed no birds have been effectively protected against agricultural intensification.[37] The Commission, Council and European Parliament have not paid enough attention to the ecological consequences of the agriculture and fisheries policies. EU policies have led to much over-grazing, causing impoverishment of poor soils, and to over-fishing. Recent reforms of the common agricultural policy, for the period 2023–2027, have been criticised by BirdLife Europe as "totally incompatible with [...] promises to transform agriculture and [...] commitments under the Climate Law and Biodiversity Strategy".[38] In turn, many of the Member States' draft national strategic plans under this inadequate policy have been criticised by the EU Agriculture Commissioner for falling short of the Commission's environmental ambition.[39] EU fisheries policy has failed to conserve fish stocks because it has been influenced by political pressure instead of scientific advice. In that regard, an Irish NGO – Friends of the Irish Environment – assisted by the international NGO ClientEarth recently argued before the High Court of Ireland that the total allowable catches fixed by the Council of the EU for 2020 were in breach of the provisions of the common fisheries policy. The High Court decided in February 2022 to refer this question of alleged illegality to the CJEU, in what will be an important test case.[40]

The Birds and Habitats Directives depend very much on the Aarhus Convention and the Environmental Impact Assessment Directive. Without them, the two Directives would have been much less successful, even if they had been supported by the Natura 2000 programme.[41] However, not enough has yet been done to ensure that the Aarhus Convention requirement that accessing justice must not be "prohibitively expensive" is fulfilled.[42]

The Birds and Habitats Directives have had at least two other huge advantages: the support of the CJEU, and the list of important areas compiled by BirdLife International. It is also important that when the Commission was considering "modernising" the Directives by weakening the protection given against economic considerations, non-governmental organisations mounted one of the largest campaigns ever seen in the EU to prevent it.[43]

EU nature conservation policy still suffers from serious weaknesses. Not enough is known about important areas in the sea and on the sea bed. Not enough weight is given to nature conservation when other EU policies are developed. Enforcement of EU law depends too much on national authorities (although this is less true of fisheries conservation). Sufficiently resourced, supported and effective national nature conservation authorities are needed. Too little has been done to prevent damage to sea-bed communities by destructive fishing gear. There are always threats from industrial and infrastructure development, and the terms of the Directives have already been weakened in this respect, by way of the Habitats Directive's amendments to the Birds Directive. Many migrating species are threatened by development outside the territories and seas of the EU.

There has been too little use of interim measures, both in national courts and in the CJEU, to protect vulnerable areas from irrevocable harm.[44]

While the final evaluation remains ongoing, the Commission's Mid-Term Review of the EU biodiversity strategy to 2020 summarises an unsatisfactory situation.[45] The new EU biodiversity strategy for 2030 sets even more demanding targets and goals, but signs of slippage are already evident, with the Commission having in March 2022 postponed its anticipated proposals for an EU Nature Restoration Law and a new Regulation on pesticides.[46] These were ultimately published in June 2022.[47]

EU nature conservation policy has benefited greatly from the fact that a great deal is known about nature in Europe as the result of the activities of voluntary organisations. Most of the information in the BirdLife International list of Important Bird Areas was gathered by amateurs, and could hardly have been collected in any other way. EU policy got the benefit of this when the lists of habitats were compiled for Natura 2000. As circumstances alter, e.g. as a result of climate change, information collected for non-governmental organisations will continue to be necessary to keep policy up to date: States cannot be relied on to do this or to provide funds for it. Adapting to climate change will make immense demands on States' resources. Even non-governmental organisations have not yet begun to think seriously about the need for translocation of organisms into suitable habitats that they cannot reach without assistance. Research has only recently begun into the consequence of longer migration routes for migrating species that may need to visit higher latitudes in summer. Non-governmental organisations will need to revise the lists of important habitats and important areas for birds in the light of climate change.

As well as the EU agricultural and fisheries policies, EU nature conservation policies need to be co-ordinated with national physical planning and land use policies, on for example quarrying, coastal reclamation, motorways, location of wind farms and other renewable energy infrastructure, and location of oil and gas development, which development must cease in order to meet net zero climate goals for 2050. In its recent "Net Zero by 2050" report, the International Energy Agency notes that "Beyond projects already committed as of 2021, there are no new oil and gas fields approved for development in our pathway, and no new coal mines or mine extensions are required."[48] The Norwegian government has carried out surveys in the Arctic to identify the most important feeding areas for cetaceans and marine birds, so as to avoid licensing gas and oil rigs in important areas. While, as noted above, such licensing needs to cease for climate reasons, elsewhere, not enough anticipatory research has been done to avoid conflicts between, for example, renewable energy infrastructure and nature conservation goals. Environmental impact assessments which should be carried out on all these kinds of developments, often have insufficient time for careful research, and the success or failure of a comprehensive nature conservation policy often depends to a great extent on the attitudes and policies of large landowners such as forestry and peat exploitation interests, and owners of large sporting estates. Areas of habitat managed for large animals such as deer, or bears are often suitable habitats for other species.

In practice, much of the success of nature conservation is due to specific localised projects and local communities. This is so in particular where conservation of the habitats necessitates management through traditional low-intensity grazing.

The Natura 2000 sites are a network for a static situation. They are not designed as stepping stones for species that may need to move their population centres (broadly, towards the northeast of Europe)[49] as a result of climate change.

As long ago as 1990, Ratcliffe warned "Even if some of the less apocalyptic predictions about the "greenhouse effect" come true, the results for wildlife and habitat… could be more serious than all the other human impacts put together."[50]

Notes

1 John Temple Lang, 'Biological Conservation and Biological Diversity', in Gunnar Sjöstedt (ed.), *International Environmental Negotiation* (Sage, International Institute for Applied Systems Analysis 1993) 171–188; Ludwig Krämer, *EU Environmental Law* (8th ed., Sweet & Maxwell 2015); John Temple Lang, 'Modern Nature Conservation' (1965) 13 Administration 200–212; Simon Lyster, *International Wildlife Law* (Cambridge University Press 1985); James A. Bailey, William Elder and Ted D. McKinney (eds.), *Readings in Wildlife Conservation* (Wildlife Society 1974); Winfried Lang, 'Les mesures commerciales au service de la protection de l'environnment' (1995) *Revue Général de Droit International Public* 545–566.
Karin Baakman, *Testing Times: The Effectiveness of Five International Biodiversity-Related Conventions* (Wolf Legal Publishers 2011).

2 Note that the European Commission recently published guidance in this area: European Commission, 'Criteria and Guidance for Protected Area Designations' (Staff Working Document) 28.1.2022, SWD(2022) 23 final.

3 Derek A. Ratcliffe, 'Nature Conservation: Aims, Methods and Achievements' (1977) 197(1126) *Proceedings of the Royal Society of London. Series B, Biological Sciences*, A Discussion on the Scientific Aspects of Nature Conservation in Great Britain 11–29. See generally Dudley Stamp, *Nature Conservation in Britain* (Collins New Naturalist Series 1969).

4 Noting, of course, that some semi-natural habitats such as hay meadows depend on human disturbance and change.

5 Hannah S. Wauchope et al, 'Protected Areas have a Mixed Impact on Waterbirds, but Management Helps' (2022) *Nature*, https://doi.org/10.1038/s41586-022-04617-0.

6 See Ian Newton, *Farming and Birds* (William Collins New Naturalist Series 2017); Graham Tucker and Michael Evans, *Habitats for Birds in Europe: A Conservation Strategy for the Wider Environment* (BirdLife Conservation Series No. 6 1997).

7 Suzanne Kingston, Veerle Heyvaert and Aleksandra Čavoški, *European Environmental Law* (Cambridge University Press 2017) ch. 12; Wouter Wils, 'The Birds Directive 15 Years Later: A Survey of the Case Law and a Comparison with the Habitats Directive' (1994) 6 *Journal of Environmental Law* 219–241; John Temple Lang, 'The European Community Directive on Bird Conservation' (1982) 22 *Biological Conservation* 11–25; John Temple Lang, 'The Wildlife Act and EEC Bird Conservation Measures', in John Blackwell and Frank Convery (eds.), *Promise and Performance: Irish Environmental Policies Analysed* (University College Dublin 1983). For the background to the Birds Directive, see Stanley Cramp, *Bird Conservation in Europe* (HM Stationery Office 1977).

8 Graham Tucker and Michael Evans, supra n 5. Graham Tucker and Melanie Heath, Birds in Europe: their conservation status (BirdLife International conservation series No. 3 1994).

9 See Paul F. Donald et al, 'Important Bird and Biodiversity Areas (IBAs): The Development and Characteristics of a Global Inventory of Key Sites for Biodiversity' (2019) 29 *Bird Conservation International* 177–198.

10 Ibid.
11 Case C-3/96, *Commission v Netherlands*, EU:C:1998:3031. See also Case C-355/90, *Commission v Spain* (Santoña marshes), EU:C:1993:331.
12 Melanie Heath and Michael Evans (eds.), *Important Bird Areas in Europe: Priority sites for conservation*, BirdLife Conservation series No. 8 (two volumes, BirdLife International 2000). A previous version of this list was Richard Grimmett and T.A. Jones, *Important Bird Areas in Europe* (International Council for Bird Preservation Technical Publication No. 9 1989).
13 "Stretching over 18% of the EU's land area and more than 8% of its marine territory [almost 27,000 sites across 27 countries], Natura 2000 is the largest coordinated network of protected areas in the world." European Commission, 'Natura 2000' https://ec.europa.eu/environment/nature/natura2000/index_en.htm (accessed 6 May 2022).
14 Heath and Evans (eds.), supra n 11.
15 See Paul F. Donald et al, supra n 8.
16 Convention on Access to Information, Public Participation in Decision-making and Access to justice in Environmental matters, Aarhus, 1998. Charles Banner (ed.), *The Aarhus Convention: A Guide for UK Lawyers* (Hart 2015).
17 Case C-77/09 *Gowan*, EU:C:2010:803. See Suzanne Kingston, Veerle Heyvaert and Aleksandra Čavoški supra n 6, pp. 94–100. See also Case C-127/02 *Waddenzee*, EU:C:2004:482.
18 European Commission, 'Communication from the Commission on the Precautionary Principle' COM (2000)1 final, 2.2.2000.
19 Case C-240/09 *Lesoochranárske*, EU:C:2011:125.
20 John Temple Lang, 'The Development by the Court of Justice of the Duties of Co-operation of National Authorities and Community Institutions under Article 10 EC' (2008) 31 *Fordham International Law Journal* 1483–1532.
21 Case C-470/16 *North East Pylon Pressure Campaign*, EU:C:2018:185.
22 The Environmental Impact Assessment Directive itself "does not lay down the substantive rules in relation to the balancing of environmental effects with other factors or prohibit the completion of projects that are liable to have negative effects on the environment" Case C-420/11 *Leth*, EU:C:2013:166. It is the combination of the principle of sincere cooperation with the environmental provisions of the Treaties and the Charter that leads to the conclusion that developments the net effects of which have been found to be seriously harmful are illegal and must not be permitted. Any other view would mean that the Treaty and Charter provisions and the Environmental Impact Assessment Directive have in the last resort no legal effects, in cases in which the environmental consequences would be seriously harmful, which could not be correct. See John Temple Lang, 'The Principle of Sincere Cooperation and the EU Nature Conservation Directives' (2021) 3 *Irish Planning and Environmental Law Journal* 114–118.
23 Charles-Hubert Born et al (eds.), *The Habitats Directive in its EU Environmental Law Context* (Routledge 2015); Andrew Jackson, *Conserving Europe's Wildlife: Law and Policy of the Natura 2000 network of Protected Areas* (Routledge 2018); Wouter Wils, 'La protection des habitats naturels en droit communautaire' (1994) *Cahiers de droit Européen* 398–430; Wouter Wils, supra n 6.
24 "A simple breakdown of globally threatened taxa in Europe by habitat-type is still not readily accessible for any group other than birds": Graham Tucker and Michael Evans, supra n 5, p. 16. The CORINE classification was primarily based on botanical criteria: Pierre Devillers and Jean Devillers-Terschuren, *A Classification of Palaearctic Habitats* (Council of Europe, Nature and Environment No. 78 1996).
25 See in particular Derek Ratcliffe (ed.), *A Nature Conservation Review Volumes 1 and 2* (Cambridge University Press 1977). "A Nature Conservative Review (NCR) is the most comprehensive nature conservation evaluation ever made in Britain, and

nothing approaching it in scope has been produced, to our knowledge, in any other country": Nature Conservancy Council (1978) Fourth Report, covering the period 1 April 1977–31 March 1978 – [1977–1978 H.C.] 646, p. 80.

26 D. Moss and B.K. Wyatt, 'The CORINE Biotopes Project: A Database for Conservation of Nature and Wildlife in the European Community' (1994) 14(4) *Applied Geography* 327–349.

27 See, for example, Jozef Keulartz, 'European Nature Conservation and Restoration Policy – Problems and Perspectives' (2009) 17(4) *Restoration Ecology* 446–450.

28 Geoffrey Griffiths and Ioannis Vogiatzakis, 'Habitat Approaches to Nature Conservation' in Andrew Millington, Mark Blumler and Udo Schickhoff (eds.), *The Sage Handbook of Biogeography* (Sage 2011) 562–571. Des Thompson, William J. Sutherland and John Birks, 'Nature Conservation and the Nature Conservation Review – A Novel Philosophical Framework' in Des Thompson, John Birks and Hilary Birks (eds.), *Nature's Conscience: The Life and Legacy of Derek Ratcliffe* (Langford 2015) 331–358.

29 Case C-371/98 *First Corporate Shipping*, EU:C:2000:600; Case C-226/08 *Stadt Papenburg*, EU:C:2010:10

30 Case C-371/98 *First Corporate Shipping*, ibid., at paragraph 22.

31 See Article 7 of the Habitat Directive, applying Articles 6(2) to (4) of the Habitats Directive in place of obligations arising under the first sentence of Article 4(4) of the Birds Directive in respect of special protection areas classified under that Directive.

32 Article 6(4) of the Habitats Directive requires that there must be an "absence of alternative solutions" simpliciter. In its guidance, the European Commission refers to "feasible" alternatives and provides detailed methodological guidance as to the consideration of such alternatives: European Commission, 'Assessment of plans and projects in relation to Natura 2000 sites – Methodological guidance on Article 6(3) and (4) of the Habitats Directive 92/43/EEC' C(2021) 6913 final (2021).

33 In the Leybucht judgment, Case C-57/89 *Commission v. Germany*, EU:C:1991:89 and the Santoña marshes case, Case C-355/90 *Commission v. Spain*, EU:C:1993:331.

34 European Commission, 'Note on Setting Conservation Objectives for Natura 2000 Sites' (2012); European Commission, 'Establishing Conservation Measures for Natura 2000 Sites: A Review of the Provisions of Article 6.1 and Their Practical Implementation in Different Member States' (2014).

35 While Article 6(1) refers to "the necessary conservation measures involving, *if need be*, appropriate management plans", the present author has argued elsewhere that management plans will in practice be needed in almost all cases, not least in order to allow for lawful appropriate assessments to be carried out pursuant to Article 6(3): see John Temple Lang, supra n 22.

36 John Temple Lang, supra n 22.

37 "Despite previous reforms, the Common Agricultural Policy (CAP) largely continues to support a resource – intensive and high-impact agricultural model which is not fit for today's societal and environmental challenges": Jabier Ruiz, 'A New EU Agricultural Policy for People and Nature' *EURACTIV* (28 April 2017).

38 BirdLife Europe, 'Press Release: European Parliament Delivers Another Major Blow to EU Green Deal: Approves Nature-Annihilating Common Agricultural Policy' (23 November 2021).

39 Natasha Foote, 'CAP Plan Drafts Fall Short of Environmental Ambitions, Says Commissioner' *EURACTIV* (23 March 2022).

40 *Friends of the Irish Environment v Minister for Agriculture, Food and the Marine, Ireland and the Attorney General* [2022] IEHC 64; see Case C-330/22 *Friends of the Irish Environment*, ECLI yet to be allocated.

41 Other EU measures that are important for nature conservation include:
Regulation 1143/2014 on invasive alien species, O.J. 2014 L 317/35.
Directive 2004/35/CE on environmental liability. O.J. 2004, L 143/56

Directive 2000/60/EC on a framework for Community action in the field of water policy, O.J. 2000 L 327/1
Directive 2008/56/EC on a framework for Community action in the field of marine environmental policy, O.J. 2008, L 164/19
The Bern Convention should be mentioned, although it is not an EU measure: Bern Convention on the conservation of European wildlife and national habitats. The EU is a party to the Bern Convention. There is also the Bonn Convention on Migrating Species, which is important for species migrating outside the European Union.

42 See James Maurici, 'Access to Justice: Review Procedures and Costs', in Charles Banner (ed.), *The Aarhus Convention: A Guide for UK Lawyers* (Hart 2015) 168–179.

43 See Chapter 2 of the present volume, by Micheál Ó Briain.

44 In the Lappel Bank Case (Case C-44/95 *R.v. Secretary of State, ex parte Royal Security for the Protection of Birds*, EU:C:1996:297), the judgment of the Court was given after the marshes had been developed as a harbour.

45 European Commission, Report: The Mid-Term Review of the EU Biodiversity Strategy to 2020, com (2015) 478, final. See also European Commission, Report: The State of Nature in the European Union, Com (2015) 219 final.

46 Gerardo Fortuna and Julia Dahm, 'EU Holds Pesticide Reduction, Biodiversity Preservation Proposals amid Ukraine War' *EURACTIV* (22 March 2022).

47 European Commission, 'Proposal for a Regulation of the European Parliament and of the Council on Nature Restoration', 22.6.2022, COM(2022) 304 final, and European Commission, 'Proposal for a Regulation of the European Parliament and of the Council on the Sustainable Use of Plant Protection Products and Amending Regulation (EU) 2021/2115', 22.6.22, COM(2022) 305 final.

48 International Energy Agency, 'Net Zero by 2050: A Roadmap for the Global Energy Sector' (4th revision, October 2021).

49 See Brian Huntley, Rhys Green, Yvonne Collingham and Stephen Willis, *A Climatic Atlas of European Breeding Birds* (Lynx 2007); James Pearce-Higgins and Rhys Green, *Birds and Climate Change: Impacts and Conservation Responses* (Cambridge University Press 2014); Peter Dunn and Anders Møller, *Effects of Climate Change on Birds* (2nd ed., Oxford University Press 2019).

50 Derek Ratcliffe, Upland Birds and their Conservation (1990) 2 *British Wildlife* 1–12.

4

ENSURING COMPLIANCE

Regulatory regimes under the EU Nature Directives

Liam Cashman[1]

Introduction

The Birds and Habitats Directives – collectively, the Nature Directives – present an impressive scientifically conceived framework for establishing and safeguarding a European Union ('EU') network of protected nature sites, Natura 2000, and periodically assessing its overall conservation effectiveness. They set out scientific processes to identify, recognise and designate the natural areas of greatest importance. They prescribe a number of general site-conservation safeguards, and define arrangements for monitoring and reporting on the conservation status of individual protected species and habitat types.[2]

However, when it comes to regulating economic activities like farming, the texts have a general character and do not address the activities directly. Furthermore, they are largely silent on compliance and enforcement mechanisms to underpin their grand design. Although lawyers helped shape and formulate them, the Directives' provisions reflect the priorities of nature conservation experts rather than those of lawyers and practitioners concerned with the enforceability and fulfilment of obligations – in particular, site-conservation ones.

In this respect, the Nature Directives fit within a wider pattern. The main instruments making up the EU environmental *acquis* (i.e. body of law) are dominated by the scientific, technical and administrative challenges of protecting air quality, water quality, waste management and nature conservation. Typically, the EU co-legislators (the Council and Parliament) have been ready to address scientific and technical challenges proactively and coherently in legislative texts but much slower to address compliance and enforcement ones. It is unsurprising that compliance and enforcement have often proved implementation's Achilles heel.

DOI: 10.4324/9780429299100-4

Leaving aside some patchy provisions[3] and some cross-cutting enforcement instruments,[4] enforcement challenges under EU environmental law have been left to the legal community (both national and EU) to sort out reactively.

In the decades since the Nature Directives were adopted, many milestones have been crossed. In 2022, the European Commission reported that terrestrial Natura 2000 sites cover 18% of the EU's land area.[5] However, assessments show that many of the habitat types and species that these are meant to safeguard have poor conservation status.[6] The dominant implementation challenge is therefore one of ensuring that Natura 2000 is managed effectively to achieve its conservation goals. Site-conservation obligations and their fulfilment are crucial to this.[7]

This chapter is compliance-focused. It explores how national authorities can ensure compliance with site conservation obligations placed on landowners, land-managers and land-users.[8] It first looks at the concepts of 'enforcement' and 'environmental compliance assurance'. It then touches on the underlying logic of the Nature Directives and the range of obligations to which they give rise, including obligations under the regulatory regimes applicable to landowners. It outlines how compliance problems and challenges can be characterised, the principles of compliance assurance, the range of interventions compliance assurance covers and strategic approaches to using them. It briefly looks at Ireland as a case-study, before drawing conclusions.

The concepts of 'enforcement' and 'environmental compliance assurance'

When they are considered at all, enforcement challenges often give rise to confusion, with distinctions blurred between different kinds of enforcement, different enforcement levels and different enforcement-related efforts.

Before proceeding, it is therefore useful to unbundle the term 'enforcement', since, depending on the context, it can carry a range of contrasting meanings.

One distinction is between 'public enforcement' and 'private enforcement', the former referring to the role of public authorities and the latter to that of individuals and environmental associations. The latter is the domain of environmental access to justice and is not treated here.

In the context of the Nature Directives, an additional distinction lies between public enforcement by the European Commission and public enforcement by competent authorities within the Member States. As guardian of the EU Treaties, the Commission has broad scope to intervene or assist in the implementation of EU laws. However, the primary responsibility for implementation lies with public authorities in the Member States.

'Public enforcement' can mean any intervention of a public authority that brings about compliance with an obligation. However, this blurs several dichotomies and several distinct means of, stages in and approaches to securing compliance.

One dichotomy is between compliance and non-compliance. Securing compliance can mean ensuring that no non-compliance occurs. It can also mean countering any non-compliance that does occur. In the first case, securing compliance has a promotional and preventive character (i.e. promoting compliance and preventing non-compliance). In the second, it has a responsive or reactive character.

When it comes to addressing non-compliance, there is a further division – between discovering the occurrence of breaches and doing something about them.

Part of the problem with the term 'enforcement' is that it is often taken to refer to punitive or coercive interventions *after* non-compliance is detected. The importance of means of detection is then overlooked or underestimated. 'Enforcement' is also sometimes equated with the imposition of penalties, which is only one type of possible – or appropriate – response.

To overcome these limitations, the European Commission has recently come to use the composite expression 'environmental compliance assurance' in place of 'public enforcement' to cover all of the cases mentioned.[9] To use a medical analogy, the term brings together preventive, diagnostic and remedial interventions.

Environmental compliance assurance thus covers three main categories of intervention:

- 'Compliance monitoring and assessment' means analysis, surveillance, inspections, investigations, audits or other interventions carried out by, on behalf of, or under the supervision of, a public authority to examine compliance by duty-holders with legal obligations;
- 'Enforcement and follow-up' means action by a public authority under civil, administrative or criminal law in response to non-compliance or suspected non-compliance with a legal obligation;
- 'Compliance promotion and prevention of non-compliance' means action to pursue compliance with legal obligations other than by means of compliance monitoring assessment or enforcement and follow-up.

The intervention logic of the Nature Directives

'Environmental compliance assurance' is a general concept and is not specifically referable to the Nature Directives. To make the connection between them, it is necessary to look at obligations that arise under the latter. However, to make sense of these obligations, it is useful to first say something about the Directives' intervention logic, drawing on the DPSIR model.

DPSIR stands for 'driver', 'pressure', 'state', 'impact' and 'response'. It is a cyclical cause-and-effect model that explains how human beings both interact with, and attempt to safeguard, the environment, in particular through public policy. It recognises that the state of the environment is influenced by pressures, such as farming practices. These pressures are themselves influenced by market and other forces called drivers – for example, to produce agricultural commodities more cheaply. The state or quality of the environment has impacts on the economy, human health and well-being, and (sometimes) people's sense of satisfaction in and responsibility for their surroundings and the natural environment. Faced with adverse impacts, policy-makers may decide to intervene. The interventions constitute a response and may be directed at drivers, pressures, state and impacts.

If we try to map the rationale and content of the Nature Directives onto DPSIR, we notice a number of things.

First, the Directives represent a response to habitat loss and species decline. These phenomena amount to aspects of a degraded state of the environment. Unlike other areas of environmental legislation, a degraded state of nature does not necessarily imply direct human health impacts (which are often invoked to justify EU legislation). Nevertheless, the degraded state has created societal impacts sufficient to generate a legislative response.

Second, the Directives place a big emphasis on the desired state of the environment. The desired state is favourable conservation status for specified habitat types and species.[10] Depending on the starting point, favourable conservation status is to be maintained or restored. Furthermore, status is to be monitored[11] and periodically reported.[12]

Third, the intervention logic of the Directives places a significant emphasis on attaining the desired state of the environment through spatial management of the pressures affecting that state, in particular via the identification and protection of the Natura 2000 site network.

Fourth, when it comes to Natura 2000, the intervention logic recognises that pressures must be addressed as regards the present, the future, and (in some situations) the past. So far as existing pressures are concerned, it is necessary to ensure that these do not result in habitat deterioration or significant species disturbance.[13] So far as new pressures are concerned, it is necessary to assess and curtail their negative impacts.[14] Finally, where past pressures have resulted in unfavourable status, it is necessary to put in place restorative measures.[15]

Addressing pressures in the dynamic way envisaged requires a system of legal obligations on those with responsibilities for the pressures. In contrast to moral obligations, legal obligations entail enforcement mechanisms because those under a legal obligation are liable to some form of enforcement action if they contravene the obligation. Depending on the obligation and the breach, different kinds of liability and enforcement may come into play.

The Nature Directives give rise to a whole lattice-work of obligations, of which an important number rest with or involve individual landowners. Before discussing the latter and their related liabilities, it is useful to address the overall system of obligations.

But there is one final point about the intervention logic: the silence of the Nature Directives when it comes to drivers – the forces that cause or magnify the pressures in the first place. This will be particularly relevant when looking later at how compliance can be promoted and non-compliance prevented.

The lattice-work of obligations

The texts of the Birds and Habitats Directives are relatively short, amounting to 20 articles in the case of the former and 24 in the case of the latter. Provisions related to Natura 2000 are even fewer in number.[16]

However, this brevity does not translate into a simple set of obligations. Rather the provisions concerned have given rise to a dense lattice-work of inter-related and inter-dependent obligations which can change – and have changed – over time.

The obligations result from different legal instruments and sources of law:

- not only the Nature Directives themselves but regulatory acts that the Directives themselves provide for[17];
- other directives[18];
- EU law in general, including primary law on the environment[19] and on fundamental rights[20];
- international law[21];
- the case-law of the Court of Justice of the European Union (CJEU);
- national primary legislation, i.e. acts adopted by Member State parliaments;
- national regulatory acts, i.e. instruments such as decrees or statutory instruments provided for under national primary legislation;
- individual consents and other individually targeted measures at national level; and
- contractual and other agreements entered into voluntarily.

Moreover, different obligations apply at different levels and concern many different actors, including the following:

- the European Commission, which has an ongoing role not only in enforcement but in implementation more generally[22];
- Member States, which have obligations vis-à-vis the European Union[23];
- individual national competent authorities, which have obligations to fulfil certain tasks, such as operate Natura 2000 consent systems for potential new pressures;
- landowners, whose activities are constrained by Natura 2000 rules and agreements put in place at national level; and
- national courts, which must determine disputes on the fulfilment of many of these Natura 2000 obligations and may sometimes need to refer to the CJEU.[24]

In addition, different obligations have applied at different periods because the establishment of Natura 2000 has been staggered in time and sometimes delayed (with the delays themselves representing breaches of obligations).

To these obligations must be added relevant legal entitlements – for example the public concerned is entitled to be consulted on site-related project decision-making and environmental associations to bring legal challenges before national courts.

A further point is that obligations may co-exist with and sometimes result from public funding.

Finally, it is important to note that the creation of some obligations is dependent on the fulfilment of others. For example, Member States must fulfil obligations to identify and recognise Natura 2000 sites in order for a system of site-conservation obligations to arise for landowners.

Regulatory regimes: site-conservation obligations on, and liabilities of, landowners

Within the overall lattice-work described, the obligations that arise for landowners can be summarised as follows:

- To respect provisions of national law aimed at avoiding site deterioration and significant species' disturbance[25];
- To respect provisions of national law to submit projects and plans likely to affect Natura 2000 sites to a consent process[26];
- To fulfil any binding commitments under conservation measures adopted by Member States to help achieve the favourable conservation status of habitat types and species.[27]

The first two clusters reflect the broad aim of Natura 2000 site conservation to protect the natural *status quo* in so far as it is important for the habitats and species that each site hosts. The obligations will typically arise under national statutory provisions, whether found in primary legislation or in regulatory acts provided for under such legislation.

In the case of the first cluster, obligations will often consist of prohibitions on carrying out physical interventions likely to harm habitats or species or both; obligations also need to cover measures to prevent natural developments that may cause the conservation status of species and habitats to deteriorate.[28]

In the case of the second, they will consist of formal procedural requirements to submit certain projects and plans for consent – and related obligations not to undertake the projects and plans without consent, or, if they are approved, outside the conditions stipulated in any consent.

The third cluster reflects an additional aim of bringing about the favourable conservation status of the individual habitats and species concerned, which may not always be attained by reliance on preventative measures alone. Here it is more likely that obligations will arise under incentive-based provisions of contracts and agreements made between public authorities and landowners than under statutory provisions.[29] The obligations will typically provide for different kinds of active land management aimed at achieving the necessary conservation outcomes.

In simple terms, the obligations described above can be divided into those having a preventive focus and those having an active management focus. Together, these correspond to the different pressure-related aspects of the Nature Directives' intervention logic. It is, of course, advisable not to be over-schematic. Preventive measures may, over time, bring about favourable conservation status as natural processes of renewal are allowed to run their course unimpeded. Likewise, conservation measures might have a preventive character as well as an active improving one.

Taken together, the obligations described can be considered to form part of a 'regulatory regime' or 'regulatory framework', i.e. a system comprising both obligations and the means of ensuring compliance with them. Those with compliance duties under the obligations ('duty-holders' or 'regulatee') constitute a 'regulated community'. The authorities responsible for overseeing and ensuring compliance with the obligations are 'regulators'. Several points are worth noting.

First, in the context of the Nature Directives, there will seldom be just one regulator who can take care of everything. Even if there is a single nature conservation authority, certain circumstances, such as the occurrence of serious nature crimes, may entail the involvement of general law enforcement authorities. Financial support schemes, in particular those depending on the EU Common Agricultural Policy, may involve a significant role for payments agencies. Furthermore, a regulatory framework can sometimes involve supervisory actors other than public authorities. Intermediaries may be important, notably in the management of contracts and agreements. It may therefore sometimes be better to speak of 'regulatory actors'.

Second, in the relationship between regulators and the regulated community, duties are not all one-sided. Regulators have obligations vis-à-vis the regulated community. Some will form part and parcel of formal consent systems – for example, regulators will be under a duty to make decisions on applications for consent to plans and projects submitted by regulatees. Others will derive from general legal requirements – for example, to respect fundamental rights or data protection requirements.

Third, regulatory regimes arising under the Nature Directives should not be considered in isolation from the wider obligations to which the instruments give rise – or indeed wider obligations under the general body of EU environmental law. This *inter alia* means taking account of views of the public concerned where consultations take place. And it means having regard to the over-arching goals of maintaining or restoring to favourable conservation status individual habitat types and species.

Fourth, not all site-conservation obligations on landowners entail the same kinds of liabilities. Breaching obligations created under statutory obligations may leave infringers liable to administrative law sanctions, such as fixed penalties. Liability may also arise under the criminal law.[30] Where the conditions for applying environmental liability rules are met, there may be civil law implications in certain situations. By contrast, for obligations created by contract and voluntary agreement, the liabilities will be typically narrower, reflecting the transactional character of the obligations.

A further set of dynamic considerations arise from the foregoing. A regulatory framework cannot be judged only in terms of its internal logic. Its success also needs to be measured according to how well it fulfils or contributes to achieving the overall goals of the Nature Directives. A regulator – and the Member State responsible for it – may be entirely satisfied with how its regulatory regime works and yet the regulatory regime may form part of a wider failure. This may be the case where a protected habitat type is at risk of disappearing or a species of becoming extinct in the Member State.

EU law and regulatory regimes

Since the Nature Directives themselves are silent on this, it is worth exploring briefly how EU law influences the content of regulatory regimes in terms of both the regulated community and the regulator.

The creation of obligations on how land is used and managed is an inevitable consequence of restraining pressures that can cause habitats to deteriorate or species to be significantly disturbed. As has been seen, restraining these pressures is part of the Nature Directives' intervention logic and is the subject of core Natura 2000 obligations set out in the Habitats Directive.

Furthermore, a teleological (i.e. purposive) interpretation of the Nature Directives requires that the extent and nature of site-conservation obligations must correspond to pressures that actually exist or could arise. If pressures on Natura 2000 are farming-related (as to a significant extent they are), then obligations must have farming relevance and must constrain potentially damaging farming activities.

Nor is it enough to have on the statute book obligations with meaningful content. There must be corresponding meaningful liabilities. This consideration relates to CJEU case-law which requires that obligations derived from EU law must be subject to sanctions which are effective, proportionate and dissuasive.[31]

Having on the statute book meaningful obligations with meaningful liabilities addresses what is expected of the regulated community. However, it is not on its own sufficient to constitute a regulatory regime since it leaves out of account the role of the regulator in overseeing compliance. This brings us to compliance monitoring and enforcement.

Since the Directives lack any provisions on compliance monitoring and enforcement, it may be wondered whether and if so how the need for such public enforcement interventions can be derived from EU law.

Ultimately, the Member State duty to monitor and enforce compliance derives from the duty of sincere co-operation that Member States owe under Article 4(3) of the Treaty on European Union. This includes a duty to enable fulfilment of obligations and a duty to refrain from jeopardising the attainment of EU objectives. The raised bog example in the Irish case-study below shows that there may be circumstances where a failure to fulfil conservation obligations and objectives can be attributed directly to a failure to monitor and enforce compliance.

The challenge of compliance assurance: characterising non-compliance

In an ideal world, compliance will follow seamlessly from the creation of obligations. However, for a multiplicity of reasons, compliance is not inevitable, and effective implementation needs to include a will and capacity of regulatory authorities to address non-compliance.

There are three different ways of characterising non-compliance. The first analyses it by reference to specific legal obligations and the conduct involved. The second looks at the factors that influence that conduct. The third looks at the footprint of non-compliance in space and time.

Each type of characterisation offers a different kind of understanding. The first is primarily relevant for knowing what to do in response to individual instances of non-compliance, i.e. breaches. The second and third are relevant for knowing what to do strategically as well as in response to breaches.

Analysis by obligation and conduct covers circumstances that point to a breach (such as the influx of a large amount of sediment into a river hosting a protected population of freshwater pearl mussel); the legal obligations likely to have been infringed (which may be both procedural and substantive); the liabilities that may arise (which may include liabilities under administrative, criminal and civil law); the parties potentially responsible and their degrees of culpability; and aspects relevant to specific envisaged types of follow-up and enforcement (for example, criminal enforcement will typically require compelling evidence).

Characterising non-compliance according to factors that influence duty-holders' conduct means looking at duty-holders' perceptions of both obligations and the control of compliance. Perceptions of obligations cover knowledge of obligations; the costs and benefits of complying with them; and the extent of acceptance of those obligations. Compliance problems may stem from ignorance, perceptions that obligations bring only burdens and no benefits, and the lack of social acceptance. Perceptions of control of compliance bring in different dimensions:

- Duty-holders' respect for authority;
- Non-governmental control, i.e. the extent to which third parties might detect and disapprove of non-compliant conduct;
- The risk of being reported;
- The risk of inspection or other checks;
- The risk of detection;

- Selectivity in inspections or other checks, i.e. the risk of being targeted;
- The risk of sanctions;
- The severity of sanctions;
- The existence of free riders, i.e. duty-holders known to act with impunity.

As can be seen, the conduct-related factors relate in part to the role of regulatory authorities. If regulatory authorities are inactive – and known to be inactive – this is likely to affect conduct. Wider societal perceptions are also relevant. If the loss of biodiversity is not part of public discourse, it may be more difficult for regulatory authorities to act, particularly if the political level is not persuaded of the need for action – or is actually averse to it.

Turning to the footprint in space, the vastness of the rural environment may mean that non-compliance is hard to detect. Sometimes effects will be known but not causes. Scale may be significant. There may even be mass non-compliance – with a large number of rural activities being carried out in disregard of Natura 2000 habitat-conservation prescriptions, for example. As for frequency and duration in time, one-off occurrences need to be distinguished from breaches that are continuing or periodically recurrent. The sensitivity and vulnerability of habitats and species are further factors, with effects either reversible or irreversible. Cumulative effects may also be important.

Compliance assurance principles

Compliance assurance is a principle-based concept. The first principle is to promote compliance and prevent non-compliance. The second is to discover and understand breaches. The third is to deal effectively with those breaches that are discovered.

The last gives rise to a number of ancillary principles:

- To end breaches as soon as possible,
- To apply sanctions effectively, dissuasively and proportionately,
- To ensure that non-compliance does not bring economic gain or advantage,
- To deter future non-compliance,
- To remediate any environmental damage.

Based on these ancillary principles, authorities may need to pursue several outcomes at the same time. For example, where they discover a serious ongoing illegal pollution discharge from a farm installation, they will need to consider intervening to stop it, sanction it and remediate the harm it has done.

The principles and the potential need for parallel interventions are explained and justified by the characterisation of non-compliance and the overall lattice-work of obligations.[32]

Organisational aspects of environmental compliance assurance

Applying environmental compliance principles involves public authorities carrying out compliance assurance interventions. There need to be organisational structures for this to happen.

The institutional and administrative arrangements for ensuring compliance with the Nature Directives vary considerably across and even within Member States. Core responsibilities include:

- *Policy direction.* Authorities with ultimate political responsibility, such as governments and individual ministers, can be expected to support and steer compliance assurance, particularly in terms of public discourse, policy-setting and budget allocation. There may be a need for strong co-operation and co-ordination, for example between agricultural and environmental ministers and their ministries.
- *Consent systems and management plans.* Authorities responsible for deciding on Natura 2000 consent applications help define and manage the overall set of applicable rules. Those responsible for management plans help create the context for prioritising conservation interventions, including compliance assurance ones.
- *Monitoring the environment and monitoring and assessing compliance.* See the next section.
- *Administering financial support.* Many farming and other rural activities in Natura 2000 benefit from public financial support. Payment agencies have a key role in compliance assurance, checking fulfilment of cross-compliance conditions, for instance. They may be amongst the principal compliance monitoring authorities.
- *Regular auditing.* See the next section.
- *Enforcement.* Monitoring authorities will sometimes have powers to impose sanctions themselves if they come across breaches – typically administrative penalties. However, use of criminal sanctions will typically require the involvement of general prosecutors.
- *Advice.* Farm advisory and similar bodies focused on supporting and advising farmers and other duty-holders may have a crucial role in compliance promotion, in particular in the preparation and dissemination of practical advice on achieving compliance under the particular conditions of individual land-holdings.

It is not enough to have authorities with nominal responsibility. It is necessary that their mandates and functions collectively cover all relevant aspects of compliance assurance. They need to be assigned duties and vested with necessary powers. They need to have the right capabilities and the necessary capacities in terms of staffing and other resources.

Within the public sector, other public authorities may exercise valuable ancillary roles. Statistical agencies may shed light on land ownership, demographics, and patterns and trends in land-use. Academic and research institutions may develop, consolidate and communicate scientific,[33] technical and legal knowledge. Farm colleges may help young farmers understand the challenges of managing biodiversity. Supreme audit bodies may provide critical oversight of the spending of public funds in Natura 2000 and the quality of public administration devoted to compliance assurance there.

Outside of the public sector, the range of potentially relevant actors is also wide. Apart from the duty-holders themselves, it can extend to intermediaries such as farm advisers and consultants; business associations, including co-operatives and farm unions; businesses dependent on Natura 2000[34] or on supply chains that originate there; certification bodies[35]; environmental NGOs focused on nature conservation or NGOs devoted to recreational interests such as angling; and members of the public, including recreational users of the countryside.

Given the range of actors, the management of relationships between them is an important facet of compliance assurance. Amongst the authorities with core responsibilities, inter-agency co-ordination and co-operation may find expression in regular strategic meetings between top officials; the conclusion of memoranda of understanding and other written agreements (such as common strategies); the use of protocols to guide interaction between officials on the ground; and the establishment of networks of practitioners. Several of the practitioner networks that exist at European level contribute to work on compliance assurance in Natura 2000.[36]

Compliance assurance interventions: monitoring and assessment

Monitoring and assessing compliance can be used to:

- determine whether duty-holders are compliant with their obligations;
- identify non-compliance and determine:
  - its nature;
  - how extensive it is;
  - what consequences it might have already had, or still have, for Natura 2000;
  - who is responsible for the non-compliance;
- determine the causes of the non-compliant conduct, including motivation (i.e. whether conduct was deliberate or accidental);
- contribute to effective enforcement and follow-up by
  - providing evidence for application of sanctions or other enforcement responses;
  - supporting responses through oversight of their implementation;
- contribute to future objective-setting for duty-holders and wider environmental management and policy-setting.

Monitoring and assessment are unlikely to be effective if they are myopic, i.e. if they only look at individual breaches in isolation. Breaches have a wider context and it is necessary to take account of 'the big picture'.

Environmental science will inform an understanding of cause and effect – for example, the precise reasons why the release of sediments or nutrients into a water body will harm aquatic species. When it comes to conservation needs and threats, Natura 2000 is characterised by many specificities, and scientific insights will frequently be highly localised in terms of the conservation interests involved. There will often be room for scientific research, since the interaction of pressures and environmental outcomes may not always be fully understood.

Taking account of the big picture also means understanding the drivers at work in the rural economy. By identifying these, regular sectoral analysis may pick up likely future pressures on land-use or habitat types within Natura 2000. For example, there may be economic drivers behind a shift from permanent pasture (with associated Natura 2000 habitat types) to tillage, or an intensification of exploitation of protected woodland. Or there may be drivers leading to land abandonment or 'hobby' farming.

Natura 2000 is subject to surveillance and reporting obligations, and the information these provide is important to compliance monitoring.[37] It can indicate the species and habitat types most at risk and the pressures to which they are subject.

State-of-environment monitoring can overlap with compliance monitoring. The territorial extent of Natura 2000 and the limited number of officials available to take care of it make it desirable to maximise the value of available information. Earth observation using aircraft and satellites can be employed to monitor the condition of land and land-use changes over large areas. Monitoring results can help to build a picture of the conservation status of habitat types such as wetlands: they can also disclose large-scale pressures such as the over-abstraction of water, which in turn may point to compliance problems.

Earth observation techniques have their limits. For example, obtaining high-resolution satellite imagery may be too expensive for the monitoring of large areas.[38] The results of earth observation will also often need to be complemented by 'ground truthing'.

The classic compliance monitoring technique is the inspection, usually involving a site-visit. While there are no explicit inspection provisions in the Nature Directives, there is a general 2001 recommendation of the European Parliament and Council on environmental inspections.[39] This embodies a widely recognised inspection model based on routine and non-routine inspections and inspection plans. Routine inspections are generally guided by risk assessment. Risk covers both the likelihood of non-compliance and its possible impacts. In a Natura 2000 context, state-of-environment, sectoral analysis and earth observation monitoring might all be used to assess risk.

Routine inspections need not necessarily focus on isolated land-holdings. They can be catchment- or area-based. For example, the Scottish Environmental Protection Agency has developed a strategy to address diffuse water pollution from agriculture, a key pressure affecting achievement of several environmental objectives, including Natura 2000 ones.[40] Since 2010, the strategy has involved officials walking the full length of the main rivers within priority catchments, recording diffuse pollution risks and non-compliance. This has been followed by a publicity campaign and visits to individual high-risk farms.

Non-routine inspections are intended to be a response to contingencies such as accidents, incidents and complaints. For example, environmental NGOs might submit credible complaints about land-use changes or likely damaging interventions – observed use of earth-moving equipment on a protected sand-dune system, for instance. Inspection plans are a means of setting out procedures for responding to such contingencies.

Complaints are an important source of information on non-compliance. Their value can be enhanced through the use of citizen science techniques, i.e. standardised processes for recording, transmitting and validating information. Complaint-sourced information can supplement other sources – putting multiple sources of information together is referred to as'geo-intelligence'.[41]

Regular auditing is closely associated with the administration of financial support. It can help in checking how effective compliance monitoring systems of compliance monitoring are in practice and in driving better overall performance of key aspects of compliance assurance. *Ad hoc* audits may be carried out by supreme audit bodies and provide similar benefits.

Interventions: enforcement and follow-up

Adherence to compliance assurance principles may become particularly challenging when breaches are detected, given the possible case for parallel or successive interventions. For example, if the breach consists in ongoing illegal land clearance in a Natura 2000 site, interventions may need to be injunctive (i.e. stopping the illegal clearance as soon as possible), punitive (i.e. sanctioning the illegal conduct) *and* restorative (i.e. remediating the ecological harm).

In practice, compliance assurance authorities will often draw on a suite of possible responses, and try to be proportionate in using them.

As noted, EU law requires that breaches be subject to sanctions that are effective, dissuasive and proportionate. In the same vein, the influential British regulatory adviser, Richard Macrory proposes that a sanctioning regime should ensure that sanctions:

- change the conduct of the offender;
- result in no financial benefit being obtained from non-compliance;
- are responsive and appropriate for the particular offender and the particular compliance issue;
- are proportionate to the nature of the offence and the harm caused;
- aim to restore the harm caused by non-compliance;
- aim to deter future non-compliance.[42]

Macrory also advises that the authorities responsible for enforcement should:

- publish an enforcement policy;
- measure the outcomes of their enforcement activities and tailor their enforcement effort to improving those outcomes;
- be able to justify their choice of enforcement actions and explain why these actions are appropriate;
- ensure that their sanctions are credible to offenders;
- be transparent in what formal enforcement activity has been taken in order to safeguard all stakeholders;
- be transparent in the methodology for calculating administrative penalties.

In a first step, an enforcement authority may use administrative tools other than fines, in particular warning letters, or statutory notices to cease and desist or carry out improvements. Authorities may also review existing consents and tighten the conditions contained in them.

Administrative fines and other financial penalties may be levied under administrative law. Or payments may be withheld if the non-compliance takes the form of non-fulfilment of conditions under a contract or agreement.

Although seldom used in practice, the Environmental Liability Directive (Directive 2004/35/EC) can be a basis for requiring remediation if non-compliance causes significant damage to biodiversity.[43]

The gravity of some breaches may point to the need to use criminal sanctions. Other factors may also come into play when deciding to bring criminal proceedings, such as repeat or persistent non-compliance, a failure to comply with statutory notices, knowingly or recklessly supplying false or misleading information, and obstruction of compliance assurance officials.

Interventions: prevention and promotion

So far as duty-holders are concerned, prevention and promotion can serve to:

- avoid drivers that lead to negative pressures (see the headage payment example in the Irish case-study later in this chapter);
- ensure that obligations are clear and understandable;
- involve duty-holders in the design of obligations or commitments that are very specific and localised;
- communicate to duty-holders the importance of compliance, in particular for protecting the environment;
- provide advice and guidance to duty-holders on how to comply with obligations;
- help duty-holders to achieve and maintain compliance as efficiently as possible, thereby reducing costs;

- recognise and reward excellence in achieving compliance.
- So far as the wider public is concerned, they can serve to:
- communicate, including through the media, the importance of compliance, in particular for protecting biodiversity;
- build public support and publicise success stories.

Good policy design and integration of environmental protection into all policy areas can help to avoid sending mixed messages to duty-holders (see again the headage payment example below).

The clarity, precision and enforceability of obligations are an obvious starting point for prevention. If obligations are unclear, there is a risk of non-compliance arising through ignorance or confusion on the part of the duty-holder. Clarity is also important for the regulatory authority, which needs to be in a position to assess whether a duty-holder is compliant.

Precision may be especially important where obligations are aimed at achieving very specific results at specific locations. For example, the conservation of individual habitat types and species under the Nature Directives may require tailored, local commitments in how farming, forestry and other forms of land-management are undertaken. One possibility is for authorities and landowners to agree, through a process of co-design, results-based measures that reflect the best scientific advice on the one hand and the insights of landowners on the other (see the Burren example in the Irish case-study below). This would appear to be particularly useful for transactional obligations.

Targeting of compliance promotion interventions entails identifying the regulated community most concerned; identifying the risks of non-compliance likely to arise through lack of awareness of obligations and the outcomes they seek to achieve; identifying the best means of reaching the target audience; and taking account of the roles that can be played by other actors such as farming associations.

Compliance promotion may take place at different levels and take many forms, including: face-to-face meetings with individual landowners; technical assistance, with provision of specialists to visit and provide tailored advice; training courses on how to undertake specific activities; demonstration farms; printed publications such as guidance documents; web-based information, in the format of Frequently Asked Questions (FAQs), for instance, or instructive videos; conferences, workshops and meetings; provision of information through media outlets, including trade media; and operation of award schemes that recognise and publicise best practice.

It is worth noting the potential contribution that permanent structural arrangements can make to prevention and promotion, for example: farm advisory services and other forms of expert intermediation; partnerships with environmental NGOs or business associations; and vocational education establishments, such as farm colleges.

The importance of data

Compliance monitoring and enforcement are both dependent on good data. Geo-intelligence implies a capacity to put different kinds of spatial data together. In principle, this capacity should be facilitated in part by the data-sharing provisions of the INSPIRE Directive.[44] Amongst other things, this instrument makes special provision for sharing of spatial data between public authorities.[45]

Strategic approaches

As the Irish case-study below shows, compliance assurance will often emerge late in the day in reaction to specific circumstances – and often as a politically unwelcome necessity. This is not entirely surprising: EU policy-makers and co-legislators have not addressed environmental compliance assurance in the same systematic structured way that they have other aspects of nature conservation.[46]

Compliance assurance is more likely to be successful if it is approached strategically. Ideally, it should be treated as an integral part of implementation and considered at the stage of overall implementation design. This would mean recognising the complementary roles of compliance monitoring, prevention, promotion and enforcement and having a principles-based approach to their deployment.

Although they do not expressly address compliance assurance, the provisions of the Nature Directives lend themselves to strategic approaches that incorporate it. The objectives of achieving favourable conservation status for individual habitats and species, the periodic assessments of conservation status and the profiling of individual Natura 2000 sites[47] all create a basis for knowing the farming and other practices that negatively or positively affect individual habitat types and species, both in general and at local level, and for assessing risks and identifying implementation priorities. To the extent that Natura 2000 involves land in private ownership, the Nature Directives entail a coherent system of landowner obligations.

Strategic approaches can involve having the right organisational structures in place; having a risk-based approach to non-compliance and its impacts; identifying priorities in terms of conservation interests and threats; looking at drivers, avoiding negative ones and creating positive ones where possible; identifying and deploying the right mix of compliance assurance interventions for individual priorities; collecting data on compliance assurance; using input and output indicators and relating these to outcome indicators (in the case of Natura 2000, the conservation status of habitat types and species and the condition of individual Natura 2000 sites).

Ireland: a case-study

To better grasp what environmental compliance assurance means in reality, it is useful to look at the experience of implementation of the Nature Directives in one Member State – touching on the roles of the European Commission and the Member State's authorities, and the inter-play between these. Ireland is a useful case-study because, prior to adoption of the Nature Directives, it had itself taken few significant initiatives in site-based nature conservation and was already experiencing serious land-use pressures on native habitat types and species, notably from diffuse water pollution, livestock overgrazing, land reclamation, agricultural intensification, commercial forestry and peat extraction.

Environmental compliance assurance is a general concept. Although developed with Member State compliance assurance in mind, it can also be used to characterise work that the European Commission does in its role as guardian of the Treaties. In this context, there is a single duty-holder and a single interlocutor, namely the Member State – here Ireland. European Commission – and wider EU – efforts to ensure compliance extend across the three categories of compliance assurance intervention. Prevention and promotion can be seen in the deployment of means such as publication of Commission guidance documents and the Commission's role in financial support, whether through the Common Agricultural Policy or LIFE, the dedicated environmental fund (which has a particular focus on supporting implementation of the Nature Directives).[48] Compliance monitoring and assessment can be seen in the way the Commission has received and investigated nature-related complaints and analysed fulfilment of a range of obligations on its own initiative and on the basis of implementation reports submitted by the Irish authorities. Enforcement is seen in the conduct of infringement procedures and the referral of nature-related cases to the CJEU. It may be mentioned in passing that the Commission is itself subject to compliance assurance in the form of audits by the European Court of Auditors. The Court has played a notable and critical role in overseeing the influence of EU spending on the implementation of the Nature Directives, for instance.[49]

With regard to the Natura 2000 regulatory regime applicable to Irish landowners, several points are worth noting.

First, farmers are by far the most important regulated community. Of Ireland's Natura 2000 terrestrial site network, 90% is privately owned and 59% is farmed. The terrestrial network amounts to over 900,000 hectares – or 13% of Ireland's terrestrial area. About 35,000 farmers are concerned – or over a quarter of the country's total.

Second, the process of establishing the regulatory regime as it now exists was not an easy one. Sites need to be legally recognised for Natura 2000 purposes in order for landowners to be bound by site-specific Natura 2000 obligations. In Ireland, site recognition proved to be particularly fraught, as there was no significant prior network of nationally protected nature sites on which to develop Natura 2000,[50] and considerable political and landowner resistance to establishing Natura 2000.[51] Indeed, the Commission needed to secure CJEU judgments in respect of site-recognition obligations under both the Habitats Directive[52] and the Birds Directive.[53] Considerable delays in establishing Natura 2000 meant delays in establishing the site protection regulatory regime. The consequences for nature conservation of these delays in establishing the regulatory regime were, nevertheless, mitigated (if only to a limited extent) by two factors. In the first place, the EU was already providing some financial support towards site conservation. In the second place, earlier Commission enforcement of other EU environmental instruments anticipated and helped to lay the basis for regulating some of the key land-use pressures on Natura 2000. For example, from 1989, the Commission was focusing on Irish rules on the impact assessment of commercial forestry, restructuring of rural land-holdings and peat extraction.[54] Likewise, Commission enforcement in relation to diffuse water pollution anticipated and helped lay a basis for beginning to address this pressure on aquatic habitats and species.[55]

Third, as elsewhere, the regulatory regime now applicable to Irish landowners is made up of parts focused on regulation of potentially damaging activities and parts focused on incentivising landowners' positive active management of Natura 2000. The former rest on a system of statutory notices addressed to individual landowners, specifying which land-use activities are prohibited, as well as consent systems governing land-use projects. Respect for certain Natura 2000-related prescriptions also forms part of cross-compliance under the Common Agricultural Policy. The incentive parts of the regime rest on a range of different financial support schemes, many (though not all) funded or co-funded by the EU.

Fourth, in organisational terms, the core responsibilities are spread across many different Irish public authorities, although one government department has overall responsibility for Natura 2000,[56] with a departmental body, the National Parks and Wildlife Service (NPWS), playing a central role. In fact, the way in which Ireland has legislated for and arranged Natura 2000 site protection means that there is a multiplicity of regulatory authorities, with major (though seldom considered) implications for compliance assurance. Depending on the project, local authorities, the Environmental Protection Agency (EPA) and other departments will not only be the consent authority but the authority responsible for monitoring compliance and taking enforcement action. Cross-compliance is a responsibility of Ireland's Department of Agriculture, Food and the Marine, and it also has a key role in administering agri-environmental financial support.

All of this would suggest the need for a strong system of co-ordination and co-operation on compliance assurance across all Irish Natura 2000 regulatory authorities. This does not exist, however, and there is no common approach to compliance assurance.[57] In other domains, notably waste, Ireland has a well-regarded model of inter-agency co-operation and co-ordination on environmental compliance assurance. This is NIECE,[58] established by the EPA, but this does not cover the Natura 2000 regulatory regime, except incidentally. Similarly, the importance of spatial information in managing Natura 2000 would suggest the need for data-sharing across regulatory authorities – for example, land-parcel information collected under the Common Agricultural Policy. In brief, there is still no fully fledged strategic approach to compliance assurance for Natura 2000 in Ireland.

The absence of a strategic approach notwithstanding, Ireland provides interesting illustrations of each category of compliance assurance intervention.

The CJEU judgment in Case C-117/00[59] highlights the complexities of establishing and implementing a regulatory regime. The Court found that, by not controlling grazing pressure by sheep over very extensive areas of commonage (i.e. areas held and used in common by multiple landowners), Ireland had amongst other things infringed the Natura 2000 requirement to prohibit habitat deterioration and significant species' disturbance. The case was instituted before Natura 2000 was fully in place, but it meant that commonage landowners in many Natura 2000 sites became subject to new requirements to de-stock sheep in order to reduce grazing pressure. The driver of the pressure had been a system of headage payments which incentivised high stocking levels without taking account of the carrying capacity of the land. Serious loss of vegetation and erosion resulted, with a high impact on some habitat types such as blanket bog and machair. Avoiding and addressing such negative drivers represents an aspect of compliance assurance. Reversing the policy signal to landowners and bringing about de-stocking meant close engagement with duty-holders and careful compliance monitoring. A particular challenge was the free-rider risk (i.e. the risk that some commonage owners would maintain high stocking levels when others reduced theirs).

When it comes to prevention and promotion, the Burren Programme[60] is now widely recognised as a positive model for collectively engaging farmers in active site management. The Burren is about 250 km^2 in extent and contains one of Europe's largest surface areas of limestone pavement. As such, it has high importance for this priority habitat type's conservation. Land in the Burren was traditionally used for pastoral farming, with cattle grazed in different parts over different periods of the year. However, in recent decades, the limestone itself came under threat, with destruction occurring to create larger fields: documented cumulative damage is cited in the CJEU judgment in Case C-392/96.[61] In some places, land abandonment emerged as a threat, with scrub encroachment reducing the Mediterranean and Alpine flora for which the area is ecologically significant. The preventive part of the Natura 2000 regime addressed the physical destruction of limestone, but the challenge of active conservation management remained. Between 2004 and 2010, a pioneering LIFE project sought to promote sustainable farming practices through close collaboration between ecologists and local farmers.[62] The approach that was developed has continued under the Burren Programme. The Programme provides farmers with results-based payments. Results correspond to the condition of habitats and species. The better the results, the higher the payments. Delivery is underpinned by a team of ecologists, demonstration farms, training, demonstration events and an awards system. This ensures a high level of farmer awareness of and sensitivity to conservation outcomes. Compliance assurance is designed into the Programme, with promotion, monitoring and responses to non-delivery very closely and logically aligned.

Turning to monitoring and enforcement, the experience with Irish raised bogs is instructive. Active raised bogs[63] represent a priority habitat type. They have largely vanished from western Europe and Ireland is especially important for the conservation of those examples that survive. The Commission recognised this from the late 1980s, helping to direct European funds for site identification and acquisition and invoking the Environmental Impact Assessment Directive to try to strengthen their protection prior to the establishment of Natura 2000.[64] Peat bogs have always been used for turbary, i.e. peat extraction for fuel. However, this pressure was greatly exacerbated by the introduction of machinery and grants for the purchase of such machinery,[65] which speeded up the rate of destruction of active raised bog. Individual peat bogs will typically contain many plots that are individually exploited, and a common pattern involves plots being exploited on behalf of multiple plot-owners by specialist contractors. Cutting and mechanical extraction were meant to stop with the advent of tighter impact assessment rules and Natura 2000 prescriptions against peat extraction. However, successive Irish governments made a deliberate political choice not to enforce the prescriptions against peat extraction, seeking to instead phase out the practice through reliance on incentives to landowners. Mechanised peat extraction continued on a wide scale, with no attempt to control it – the unremitting loss of areas of bog of nature conservation importance already noted by the CJEU in 1999[66] simply continuing as before. By 2010, status-monitoring of the habitat type indicated that 35% had been destroyed since 1992, the annual rate of loss being between 1 and 4%. With the habitat type facing an existential threat, the Commission launched an infringement procedure at the beginning of 2011 – enforcement in respect of non-enforcement, as it were.

Now itself under pressure, the Irish government began to enforce the peat-cutting prohibition, but in many places it met with organised landowner defiance, leading to public disorder and the involvement of Ireland's police force. In these fraught circumstances, the Irish authorities deployed a combination of the classic compliance assurance interventions this chapter describes: monitoring compliance on individual sites, including through earth observation; issuing warnings and instituting a number of criminal prosecutions[67]; and intensifying efforts to engage with the regulated community. Positive engagement meant dialogue with landowners and their representatives. It included addressing a serious issue of social acceptance of obligations relating to peat extraction (i.e. the aspect of duty-holder perceptions of obligations mentioned earlier) and addressing landowner grievances centred on obtaining satisfactory financial compensation and/or being allowed to continue to extract peat (with the feasibility of providing less sensitive sites as alternatives being explored). Significant new resources devoted to peat-bog restoration meant that local communities could begin to see economic benefits resulting from conservation.[68]

Despite these compliance assurance interventions, which have no doubt lessened the overall extent of the problem, the latest publicly available data indicate that the issue nevertheless remains to be resolved: information obtained by the Irish Wildlife Trust under access to information on the environment rules revealed that 18 of Ireland's 57 raised bog SACs were subject to peat extraction in

2021.[69] Allowing for obvious differences of scale, the admission that, decades into the application of the Habitats Directive, almost a third of Ireland's most important remaining deep-peat ecosystems continued to be subject to illegal extraction invites comparison with the more publicised depredations associated with tropical rainforests. In September 2022, the Commission issued a further formal warning that it would bring the matter before the Court of Justice unless Ireland halted illegal extraction.[70] As part of a multi-phase review, a government-commissioned independent report on the NPWS, finalised in October 2021,[71] notes amongst other things a 70% cut in funding that occurred between 2008 and 2011 and a far weaker enforcement record compared with a similar-sized Irish regulatory body, Inland Fisheries Ireland.[72] The report recommends that a wildlife crime unit in the NPWS should give early consideration to the development and publication of a compliance and enforcement policy,[73] which would set out the NPWS's general approach to compliance and enforcement (this would be consistent with the strategic approaches discussed earlier in this chapter).[74] The next phase of the review – in effect, a filtering of the report – also sets out recommendations, including an NPWS re-organisation, with a new directorate to be responsible for enforcement amongst other things.[75] In this second phase, however, and in the strategic NPWS action plan which followed, the development of a compliance and enforcement policy is not mentioned, and compliance assurance is not a salient theme at all. Given the persistent pattern of non-existent or weak NPWS enforcement since the 1990s, it is hard to discern a basis for optimism concerning Ireland's political and institutional willingness to take effective action.

Conclusions

The Nature Directives, and Natura 2000 in particular, involve the creation of regulatory regimes for landowners. Achievement of conservation objectives depends on these regimes being fit for purpose and on landowners being compliant with them. The regimes must not only regulate potentially harmful activities but also incentivise beneficial ones. They therefore need to encompass very different kinds of obligation – prohibitions on causing pollution and the destruction of landscape features, on the one hand, and commitments to farm in nature-friendly ways, carry out beneficial activities or execute particular conservation works, on the other. These obligations will be found in very different kinds of legal instrument – extending from those of a legislative or statutory character, including notices, to those of a contractual or voluntary nature (which are essentially transactional). There may be corresponding variations in the regulators or regulatory actors responsible for overseeing them.

As the Irish case-study shows, compliance assurance is relevant for all these different aspects of regulatory regimes. Whatever the nature of the obligation, regulators and regulatory actors need to have systems in place to verify compliance and understand and respond to non-compliance. There are many different possibilities and choices, but all the different categories of compliance assurance intervention have a role to play. Compliance assurance is more likely to be effective if this is recognised and if regulators and regulatory actors adopt strategic approaches to using their available (and usually limited) resources and have clear policies and protocols to guide the use of different interventions.

Notes

1 The author is a retired European Commission official, having worked in DG Environment from 1988 to 2021. The content draws on work done in 2018–2019 to prepare guidance on environmental compliance assurance in rural areas. Led by the Commission, this involved experts from Member States and European networks of environmental agencies and compliance practitioners, and the Environmental Compliance and Governance Forum, a high-level expert group created by the Commission in 2018. The work culminated in the endorsement by the Forum in February 2020 of the 'Vade Mecum on environmental compliance assurance in rural areas'. This document and a related summary guide are accessible at https://ec.europa.eu/environment/legal/compliance_en.htm. The author would like to acknowledge the many insights provided by Miroslav Angelov and Andrew Farmer, as well as their roles of in preparing the Vade Mecum. Views expressed in this chapter are entirely personal. They do not necessarily represent those of the European Commission.
2 The Nature Directives include important species-protection obligations, but this chapter focuses on those relating to Natura 2000.
3 For example, recent environmental directives, although not the Nature Directives, include a standard penalty clause which requires Member States to ensure that infringements are subject to penalties that are effective, proportionate and dissuasive. See, for example Article 23 of the Water Framework Directive (Directive 2000/60/EC).
4 Directive 2008/99/EC of the European Parliament and of the Council of 19 November 2008 on the protection of the environment through criminal law, OJ L 328, 6.12.2008, pp. 28–37; Directive 2004/35/CE of the European Parliament and of the Council of 21 April 2004 on environmental liability with regard to the prevention and remedying of environmental damage, OJ L 143, 30.4.2004, pp. 56–75.
5 European Commission, Natura 2000 – Nature and Biodiversity Newsletter, February 2022, Number 51.
6 European Environment Agency, State of nature in the EU: results from reporting under the Nature Directives 2013–2018, EEA Report No 10/2020, August 2020.
7 The 2017 Action Plan for nature, people and the economy, COM(2017) 198 final, stresses the importance of building political ownership and strengthening compliance, while the EU Biodiversity Strategy for 2030, COM/2020/380 final, contains a section on 'Stepping up implementation and enforcement of EU environmental legislation', which includes striving to improve compliance assurance amongst its goals.
8 For ease of reference, a single category, landowners, will be referred to in the chapter.
9 European Commission, Commission Communication on EU actions to improve environmental compliance governance, COM(2018)10 final, and associated staff working document SWD(2018)10 final.
10 See Article 2(2) of the Habitats Directive and the definition of conservation status in Article 1(i). The Birds Directive is an earlier instrument, but its approach is broadly analogous.
11 See Article 11, Habitats Directive.
12 See Article 17, Habitats Directive.
13 Article 6(2), Habitats Directive.

14 Article 6(3) and (4), Habitats Directive.
15 Article 6(1), Habitats Directive. See also reference to 'restore' in Article 2(2) of the Habitats Directive.
16 Essentially, Article 6 Habitats Directive, although in some circumstances, notably a failure to recognise sites requiring recognition, other provisions may apply.
17 Article 4 of the Habitats Directive provides for Commission site-listing decisions. One such listing decision featured in the validity reference in Case C-281/16 *Vereniging Hoekschewaards Landschap*, EU:C:2017:774.
18 For example, the Environmental Crime Directive (Directive 2008/99/EC) requires criminal sanctions for certain offences linked, inter alia, to the Nature Directives.
19 For example, the principles set out in Article 191(2) of the Treaty on the Functioning of the European Union (TFEU) may have a role in interpretation of the Nature Directives.
20 The Charter of Fundamental Rights will be relevant to rights derived from or linked to the Nature Directives.
21 For example, the Aarhus Convention on access to information, public participation in decision-making and access to justice in environmental matters has significant implications for implementation of the Habitats Directive.
22 For example, in relation to the listing of sites.
23 Apart from transposing the Nature Directives and fulfilling obligations at national level, Member States must communicate measures and report on an ongoing basis to the European Commission.
24 See for example several preliminary references resulting from litigation initiated by Peter Sweetman (C-258/11, C-323/17, C-164/17).
25 Article 6(2), Habitats Directive.
26 Article 6(3), Habitats Directive.
27 Article 6(1), Habitats Directive.
28 See at para. 34 of Case C-6/04 *Commission v United Kingdom*, EU:C:2005:626, where the Court stated:

> As the Advocate General has observed in point 19 of her Opinion, it is clear that, in implementing Article 6(2) of the Habitats Directive, it may be necessary to adopt both measures intended to avoid external man-caused impairment and disturbance and measures to prevent natural developments that may cause the conservation status of species and habitats in SACs to deteriorate.

29 Of course, public authorities will need to have the necessary legal powers under statute to conclude contracts and agreements.
30 Under the Environmental Crime Directive (Directive 2008/99/EC), Member States are obliged to create criminal sanctions relevant to breaches of the Nature Directives. This is also the case under the proposed replacement directive: European Commission, Proposal for a Directive of the European Parliament and the Council on the protection of the environment through criminal law and replacing Directive 2008/99/EC, COM(2021) 851 final, Brussels, 15.12.21.
31 See for example para. 46, Case C-354/99 *Commission v Ireland*, EU:C:2001:550.
32 The principles of environmental compliance assurance find an echo in the principles governing EU environmental policy set out in Article 191(2) of the TFEU, for example the principles of preventive action and that the polluter should pay.
33 Both the natural and social sciences.
34 For example, angling, walking and cycling tourism interests.
35 For example, forest certification.
36 The European Network of the Heads of Environment Protection Agencies (NEPA), in particular its Better Regulation Interest Group (BRIG), takes a close interest in environmental compliance assurance, as do the following networks: the European Union Network for Implementation and Enforcement of Environmental Law (IMPEL),

which is made up of environmental authorities and environmental inspectorates at central or regional level from all EU countries as well as candidate, potential candidate and European Free Trade Association (EFTA) countries; the European Network of Prosecutors for the Environment (ENPE), which represents authorities responsible for criminal prosecutions and, to some extent, administrative sanctions; and the EU Forum of Judges for the Environment (EUFJE), which brings together judges from across the EU responsible for criminal, administrative and/or civil cases.

37 See the raised bog example in the Irish case-study.

38 An effective strategy may involve using high-resolution imagery for 'hot spots' identified as presenting a higher risk of non-compliance.

39 Recommendation 2001/331/EC of the European Parliament and of the Council of 4 April 2001 providing for minimum criteria for environmental inspections in the Member States, OJL 118, 27.4.2001.

40 Post-Brexit, Natura 2000 sites are known as 'European sites' in Scotland.

41 See judgment in Case C-441/17 *Commission v Poland,* EU:C:2018:255, and associated interim orders. In this case, which concerned logging pressures on the Puszcza Białowieska Natura 2000 site, the Commission drew on earth observation data as well as data recorded on mobile devices and submitted by environmental activists.

42 Richard B. Macrory, 'Regulatory Justice: Making Sanctions Effective', Final Report, November 2006, https://webarchive.nationalarchives.gov.uk/20121205164501/http:/www.bis.gov.uk/files/file44593.pdf

43 The Environmental Liability Directive (Directive 2004/35/EC) focuses on preventing and remediating 'environmental damage', which includes 'damage to protected species and natural habitats', a category closely linked to the Nature Directives. The importance of the link between the Environmental Liability Directive and the Nature Directives can be seen in Case C-297/19 *Naturschutzbund Deutschland – Landesverband Schleswig Holstein e. V,* which concerns adverse effects on a Natura 2000 site arising from maintenance of land drainage for agriculture. An amendment to the Environmental Liability Directive adopted in 2019 required the European Commission to develop guidelines providing a common understanding of the term 'environmental damage'. These guidelines were published in 2021: European Commission, Commission Notice: Guidelines providing a common understanding of the term 'environmental damage' as defined in Article 2 of Directive 2004/35/EC of the European Parliament and of the Council on environmental liability with regard to the prevention and remedying of environmental damage, OJ C 118, 7.4.2021, pp. 1–49.

44 Directive 2007/2/EC of the European Parliament and of the Council of 14 March 2007 establishing an Infrastructure for Spatial Information in the European Community (INSPIRE), OJ L 108, 25.4.2007, pp. 1–14.

45 Article 17.

46 The Commission's 2018 communication and related actions are an attempt to address this gap – supra n 9.

47 Through the Natura 2000 site forms provided for in Article 4 of the Habitats Directive.

48 The LIFE instrument has been periodically renewed. The current one is Regulation (EU) 2021/783 of the European Parliament and of the Council of 29 April 2021 establishing a Programme for the Environment and Climate Action (LIFE), and repealing Regulation (EU) No 1293/2013, OJ L 172, 17.5.2021, pp. 53–78.

49 See, for example, European Court of Auditors, 'More efforts needed to implement the Natura 2000 network to its full potential', Special report 01/2017.

50 While there were protected nature sites, these were almost all in state ownership and their territorial coverage was very low.

51 See, for example, Brigid Laffan and Jane O'Mahony, 'Bringing Politics Back In'. Domestic Conflict and the Negotiated Implementation of EU Nature Conservation Legislation in Ireland' (2008) 10(2) *Journal of Environmental Policy & Planning* 175–197.

52 Case C-67/99 *Commission v Ireland,* EU:C:2001:432.
53 Case C-418/04 *Commission v Ireland,* EU:C:2007:780.
54 Case C-392/96 *Commission v Ireland,* EU:C:1999:431.
55 See Case C-396/01 *Commission v Ireland,* EU: C:2004:136, which addressed Ireland's poor implementation of the Nitrates Directive (Directive 91/676/EEC) and Case C-282/02, *Commission v Ireland,* EU: C: 2005: 334, which addressed Ireland's poor implementation of the Dangerous Substances Directive (Directive 76/464/EEC). These cases put a focus on eutrophication, one of Ireland's main water quality problems. In practice, diffuse water pollution continues to be a significant environmental issue in Ireland, as can be seen from successive EPA state-of-the-environment reports. In July 2022 it was reported that an environmental NGO will later in the year be seeking leave from the High Court to judicially review Ireland's approval of its most recent Nitrates Action Programme on environmental grounds: Aodhan O'Faolain, 'An Taisce Challenges State Nitrates Plan Over Concern for Waterways', *The Irish Times*, 20 July 2022.
56 Department of Housing, Local Government and Heritage.
57 The 2021 report on the NPWS notes that, in 2021, it entered into a joint protocol with Ireland's police force in relation to wildlife crime. It is not clear, however, that this extends to crimes concerning Natura 2000 – and in any case it does not cover all of the regulatory authorities with Natura 2000 responsibilities.
58 Network for Ireland's Environmental Compliance and Enforcement.
59 Case C-117/00 *Commission v Ireland*, EU:C:2002:366.
60 www.burrenprogramme.com
61 C-392/96, *Commission v Ireland*, para. 80, EU:C:1999:431.
62 BurrenLIFE – Farming for conservation in the Burren LIFE04 NAT/IE/000125.
63 i.e. still peat-forming.
64 See Case C-392/96. The EIA Directive requires environmental impact assessment of environmentally significant peat extraction projects.
65 Case C-392/96 provides a site-specific example of the use of peat extraction machinery, see para. 36.
66 Case C-392/96, para. 78.
67 These do not appear to have resulted in any convictions. The author remains unaware of any sanctions having ever been imposed for illegal peat extraction in an Irish Natura 2000 site.
68 For example, the LIFE-funded project 'The Living Bog' (2016–2020), aimed at restoring 12 raised bog SACs (http://raisedbogs.ie/).
69 Irish Wildlife Trust, 'As the government heads to the COP26 Climate Summit in Glasgow, new information shows that unlawful turf-cutting remains widespread on Special Areas of Conservation', 28 October 2021.
70 Tim O'Brien, 'EU Gives Ireland Two Months to Halt Peat-Cutting in Conservation Areas or Face Court', *The Irish Times*, 29 September 2022.
71 Jane C. Stout and Micheál Ó Cinnéide, 'A Report to the National Parks and Wildlife Service, of the Department of Housing, Local Government and Heritage', October 2021.
72 Ibid., section 4.12.
73 Ibid., Recommendation 22.
74 A qualification is that Recommendation 22 does not advert to the importance of compliance assurance principles and may be interpreted as only looking to criminal-law enforcement. Successful criminal prosecutions have a role to play in sanctioning and deterring but may be ineffective in halting ongoing damage (especially if subject to low penalty caps) or remediating damage already sustained. Compliance assurance principles point to the need for injunctions and remediation notices as well.
75 Gerry Kearney, 'Reflect and Renew – A Review of the National Parks and Wildlife Service', February 2022.

5

THE MARGIN OF APPRECIATION WHEN APPLYING ARTICLE 6(3) AND (4) OF THE HABITATS DIRECTIVE

Christoph Sobotta

Introduction

In principle, we can understand law and judicial review as a mathematical model. Rules are algorithms and therefore their application depends on calculations. If we frame this model in a very simplistic (or idealistic) way, there can only be one correct outcome. Either the sums add up or they do not. Courts would simply need to verify this. However, reality is significantly more complex. For many cases, we do not know all the numbers and calculations necessary to apply the rules in this way. Nevertheless, decisions need to be taken and therefore legal systems allow for a margin of appreciation that is not subject to a strict mathematical review. In mathematical terms, fuzzy logic might be appropriate.

We will find many examples of this approach in EU environmental law if complex assessments and/or the balancing of conflicting interests become necessary.[1] Therefore, the margin of appreciation that authorities enjoy in this regard and the standard of judicial scrutiny courts apply are of particular interest. The jurisprudence of the Court of Justice of the European Union (CJEU or Court) on this topic may appear ambivalent because, on the one hand, in most instances, the Court exhibits a large degree of judicial self-restraint, but, on the other hand, with respect to Article 6(3) of the Habitats Directive[2] it applies a very strict standard. Nevertheless, on closer inspection, the ambivalence disappears because Article 6(3) and (4) taken together allow for an implementation of the more general jurisprudence of the Court. However, to arrive at this outcome two preliminary steps need to be taken: First, I will address the need for complex assessments and balancing in environmental decision-making and then I will describe the general jurisprudence of the Court on judicial review of such decisions.

DOI: 10.4324/9780429299100-5

Complex assessments and balancing

The application of environmental law, such as the rules on site protection under the Habitats Directive, often requires complex assessments. This complexity results from scientific uncertainty in relation to many questions that authorities need to answer in this context. The uncertainty may be an expression of insufficient knowledge about the effects of certain actions, but it can also be a consequence of the complexity of the affected environment.[3] Though we can eliminate the first type of uncertainty with additional research, it may not be possible to overcome the second type. Complete knowledge of complex systems mostly cannot be acquired and therefore they often react unpredictably to inputs. Returning to mathematics, we could strike a parallel to chaos theory.

In the field of the environment, EU regulatory measures, such as the ban on the additive methylcyclopentadienyl manganese tricarbonyl (MMT) in automobile fuels,[4] can serve as examples. When the EU introduced this ban, the European Parliament and Council considered that the use of MMT was very damaging to the environment. Conversely, this ban had not been in the Commission proposal because it considered the scientific evidence to be insufficient. Moreover, several years later, before the Court, the producer still argued that there was no proof of damage, and this was not definitively proven wrong by the institutions. Effectively, there was some evidence pointing towards harm, but not enough to establish this harm with certainty.[5]

The MMT ban also illustrates another element of environmental decision-making, namely the balancing of competing interests. After arriving at the conclusion that the additive posed a risk to the environment, the legislator needed to decide whether the risk justified restricting the economic freedom of additive producers and the freedom of choice of consumers. In particular, the producer argued that the use of the additive was necessary to prevent harm to engines. Therefore, the legislator needed to consider whether less invasive measures could be adopted and whether the advantages of a ban would outweigh possible negative consequences. It should be noted that this final balancing exercise required a comparison between two sets of uncertain harm – on the one hand, uncertain harm to the environment and on the other hand highly likely economic harm to the producer, but also uncertain harm to the consumer.

Judicial review of complex assessments and balancing

Courts need to address complex assessments and balancing. Under the rule of law, effective judicial protection needs to be available and this right is enshrined in Article 47 of the EU Charter of Fundamental Rights. This protection cannot exclude decisions taken in the light of scientific uncertainty and/or difficult balancing exercises. Nevertheless, judicial scrutiny of such decisions is a sensitive matter. Should courts apply very strict standards or act with restraint? To put it more practically against the backdrop of our example: How should courts verify whether the MMT ban was justified?

Procedural autonomy of Member States

The starting point for any debate on this issue with regard to the application of EU law and its review within the Member State legal systems is the principle of procedural autonomy. In the absence of detailed EU rules governing judicial review of decisions implementing EU environmental law, it is generally for the domestic legal system of each Member State to determine those requirements, provided, however, that the requirements are not less favourable than those governing similar domestic situations (principle of equivalence) and that they do not render impossible in practice or excessively difficult the exercise of rights conferred by the EU legal order (principle of effectiveness).[6]

It is conceivable that, in exercising procedural autonomy, Member States will grant their courts very far-reaching powers of review. For example, courts could be authorised to re-evaluate the scientific assumptions supporting a decision, examine additional facts or scientific theories or rebalance the competing interests, thereby substituting themselves for the administrative authorities and modifying or substituting their decisions. In considering a restrictive rule, such as the MMT ban, we could imagine that courts might annul such a rule if they were not convinced that it was justified. It appears that in some areas the Swedish[7] and Finnish[8] legal systems go a long way in this direction by employing expert judges and German courts often examine the technical assessments of the administration very intensively by hearing court-appointed experts.[9] From the perspective of general EU law, there can, in principle,[10] be no objection to such a standard of review if it satisfies certain requirements, such as having an adequate scientific basis and complying with minimum procedural requirements resulting from fundamental principles of EU law.[11] Nevertheless, other legal systems apply more limited standards of judicial review,[12] though the Aarhus Convention[13] Compliance Committee is currently assessing whether under Article 9 of the Convention a stricter standard of review needs to be applied in environmental cases,[14] and indeed the Compliance Committee already held in 2021, in a separate case relating to Environmental Impact Assessment,[15] that:

> 119. [...] article 9(2) [of the Aarhus Convention] requires that a reviewing body must review all the facts, evidence and arguments before it and, based on that review, determine whether the contested decision is lawful. This requires the court to carry out its own assessment, in the light of all the evidence before it, as to whether the applicable legal requirements were met. The court must also clearly set out its reasoning when doing so.
>
> 120. For example, in a challenge to the substantive legality of an EIA screening decision, the court must make its own assessment, based on all the evidence put before it, as to whether the proposed activity was likely to have a significant effect on the environment and thus to require an EIA. It would not be sufficient to merely check that the decision-maker carried out the correct procedural steps for determining whether a project was likely

> to have significant effects. Nor does it suffice for the court to check that the decision-maker had formally applied the correct legal test and that the decision-maker had convinced itself that that test was met in a particular case.
>
> 121. To be clear, the Convention does not require the court to undertake a completely fresh analysis of all matters arising in the case and to substitute its decision for the decision taken by the competent authority. Nevertheless, the court must undertake its own assessment of all the evidence before it to determine whether the applicable legal requirements were met. The Committee considers that this requires the court to perform a review function over findings of fact and the weight to be given to evidence where those may have a direct impact on the determination as to whether the applicable legal test (for example, likely significant effects) has been met.

The impacts of this relatively recent finding may not yet be fully reflected in the legal systems of parties to the Aarhus Convention.

EU minimum requirements for judicial review

Conversely, the review criteria applied by the EU courts in assessing such measures adopted by the EU institutions, for example the MMT ban, allow identifying the minimum level of scrutiny, which the review of such a decision by the national courts must achieve. While EU law, in particular Article 19 of the Treaty on European Union and Article 47 of the Charter, requires that the Member States establish a procedure for effective judicial review of national decisions applying EU law,[16] this review, in principle, need not be more extensive than that carried out by EU courts in similar cases.[17]

Review of EU measures by EU courts is characterised by the broad discretion that the competent EU institutions enjoy with regard to complex scientific or technical assessments and the balancing of competing interests. The discretion relates in particular to the assessment of the facts in order to determine the nature and scope of the measures. In this regard, substantive review by the EU judicature only verifies whether the authorities have manifestly exceeded the limits of their discretion.[18]

Although the application of this jurisprudence to environmental cases could be affected, at least indirectly, by the pending case on judicial scrutiny in the Aarhus Convention Compliance Committee as well as by the Committee's recent decision cited above,[19] this jurisprudence is not merely an example of judicial self-restraint, but a practical expression of institutional balance. It is based on the consideration that EU courts cannot substitute their assessment of scientific and technical facts for that of the institutions upon whom alone the legislature has placed that task.[20] These other institutions possess greater technical and scientific expertise within the fields of their competence than courts of law. Therefore, they are better placed to conduct complex scientific assessments, such as the risk assessment of MMT. In addition, such institutions usually enjoy a stronger democratic legitimacy than the courts because they are directly or indirectly responsible to an elected parliament.[21] This legitimacy justifies the balancing of competing interests. In the example of the MMT ban, the initiative for the ban actually came directly from the European Parliament and it was taken up by the Council, that is by Member State governments responsible to their parliaments. In the absence of democratic supervision, as can be the case for specialised independent bodies such as the European Central Bank,[22] there is at the very least a democratic decision attributing a specific margin of appreciation to the body in question and this can be reversed by another democratic decision.

Conversely, the exercise of discretion comes with other obligations that are subject to stricter scrutiny. Specifically, the competent institution is under the duty to examine carefully and impartially all the relevant aspects of the individual case[23] and to comply with the relevant procedural rules.[24] Essential components of the latter are specified by the right to good administration as laid down in Article 41 of the Charter. While Article 41 of the Charter only applies to EU institutions, Member State authorities applying EU law are subject to similar obligations resulting from an unwritten general principle of EU law.[25] These obligations include in particular the right of every person to be heard, before any individual measure that would affect him or her adversely is taken[26] and the obligation on the part of the administration to give reasons for its decisions. These reasons allow the person concerned and eventually the courts to consider the legality of the measure,[27] and in particular, whether the authorities have accurately examined the relevant facts. Therefore, a statement of these facts and of their consideration should be part of the reasons given.

Compliance with these obligations is essential to the proper exercise of the discretion granted to the authorities. Irrespective of their technical competence and their democratic legitimacy, a decision of high quality requires the investigation of all relevant facts. Hearing the parties concerned helps to identify such facts because these parties often have specific knowledge of the situation on the ground and an interest to make it known. In the case of the MMT ban, the Court therefore considered it essential that the European Parliament and Council were able to point to studies that they relied on to arrive at their assessment of the risks and that in particular, the Parliament had conducted hearings with all stakeholders on this issue.[28] With regard to decisions affecting the environment, EU legislation[29] and international agreements[30] have even extended the right to be heard into rights to public participation.

Nevertheless, even with regard to these obligations, judicial review will sometimes be limited. While courts can scrutinise strictly whether relevant parties had the opportunity to participate in the procedure or whether the competent institution gave sufficient reasons to explain a decision, in particular the question of whether the institution examined all relevant aspects is more difficult to consider. The establishment of specific facts and their relevance as well as scientific theories can represent other areas of scientific uncertainty. For example, with regard to the MMT ban, there was some dispute about whether additional studies on the effects of MMT were necessary. Therefore, the Court recognised that the discretion of the institution in question also relates, to a certain extent, to the finding of facts underlying their action.[31] Nevertheless, I would submit that the EU courts have not yet fully explored how far this discretion extends and that it may not be sufficient to apply the manifest error test in this regard. The courts should rather verify whether the institution undertook a reasonable, i.e. *proportional*, effort to identify the relevant facts and theories. The jurisprudence on the application of the precautionary principle, requiring, first, identification of the potentially negative consequences for health (or the environment) of the practice under examination, and, second, a comprehensive assessment of the risk to health based on the most reliable scientific data available and the most recent results of international research, supports this view.[32] Otherwise, arbitrary outcomes could be justified by careful selection of 'relevant' facts.

Finally, one more dimension of judicial review needs to be mentioned because, in certain cases, the substantive discretion is limited and therefore courts must examine such measures more intensively. Specifically, the Court has ruled that, where interferences with fundamental rights are at issue, the extent of the EU legislature's discretion may prove to be limited, in the light of the principle of proportionality. This limitation depends on a number of factors, including, in particular, the area concerned, the nature of the right at issue guaranteed by the Charter, the nature and seriousness of the interference and the object pursued by the interference.[33] Thus, in view of the important role played by the protection of personal data in the light of the fundamental right to respect for private life and the extent and seriousness of the interference with that right caused by a comprehensive retention of call data, Directive 2006/24[34] was subject to strict review[35] and the Court ultimately declared it invalid. Apparently, this type of issue falls between the general category of issues where the EU legislator enjoys wide discretion and the limited category of issues, such as the death penalty, prohibited by Article 2 of the Charter, that are explicitly put beyond its powers. In this interim category, the EU legislator can, in principle, still act, but the Court will not limit review to the manifest error test. Conversely, most regulatory issues in the field of the environment only result in restrictions to weaker guarantees, such as the right to conduct an economic activity under Article 16 of the Charter and the right to make use of property under Article 17 and they have not yet triggered strict review by the EU courts.[36] In addition, the objective of the measure can also be relevant in this regard. Up to now, the Court has not considered that the general objective of protecting the environment warrants strict review.[37] However, with regard to specific environmental objectives, for example, the protection of human health and human life, stricter judicial scrutiny might be indicated.[38]

Appropriate assessment – Article 6(3) of the Habitats Directive

Approaching the topic of site protection with this background in mind, one would expect that Member State authorities would enjoy wide discretion and judicial review would be limited to manifest errors. However, the jurisprudence of the CJEU, in particular on Article 6(3) of the Habitats Directive, limits Member State powers much more strictly.[39] Therefore, it appears difficult to fit the standards applied in the context of Article 6(3) into the general framework laid out above. Nevertheless, on closer examination and in the light of Article 6(4) the perceived contradictions disappear.

First sentence: assessment

Under the first sentence of Article 6(3) of the Habitats Directive, plans or projects not directly connected with or necessary to the management of the site but likely to have a significant effect thereon, either individually or in combination with other plans or projects, shall be subject to appropriate assessment of their implications for the site in view of the site's conservation objectives.

This provision, as such, sets up a procedural requirement, namely the appropriate assessment, that aims to ensure that all the relevant aspects of the case are examined carefully and impartially. In particular, the assessment procedure is intended to ensure, by means of a prior examination, that a plan or project not directly connected with or necessary to the management of the site concerned, but likely to have a significant effect on it, is authorised only to the extent that it will not (in fact) adversely affect the integrity of that site.[40] Therefore, the Court has found that where a plan or project is likely to undermine the protected site's conservation objectives, it must be considered to be likely to have a significant effect on that site[41] and, as a consequence, be subject to the assessment. More specifically, there must be a probability or a risk of such an effect.[42]

In principle, the identification of such a probability or risk requires a complex assessment. Many effects of activities on protected habitats or species are simply not known with sufficient certainty and/or detail because we do not know enough about the species and ecosystems concerned. Moreover, ecosystems in particular will often be characterised by substantial inherent complexity, thereby rendering the effects of intervention unpredictable.[43] Such uncertainty, therefore, would appear to imply wide discretion. This would be in line with the jurisprudence to the effect that the discretion also relates, to a certain extent, to the finding of facts underlying the action.

However, the Court reads Article 6(3) in the light of the precautionary principle and therefore has found that such a risk exists if it cannot be excluded on the basis of objective information that the plan or project will have significant effects on the site concerned.[44] In particular, future mitigation measures that are foreseen for the implementation of a project are not accepted as objective measures that can exclude significant effects.[45] This limitation of discretion confirms the thesis that discretion with regard to the finding of facts underlying an action is more limited than the discretion allowed with regard to other types of assessment or balancing. In the light of the purpose of the appropriate assessment, this restriction of discretion is particularly convincing because without the assessment a well-founded consideration of possible impacts on the site is difficult to imagine. Moreover, the reference of the Court to the precautionary principle that, according to Article 191 TFEU, guides EU environmental policy is an important reminder that the precautionary purpose of EU environmental law needs to be taken into account in its application and interpretation. The consequence of this jurisprudence is that, as a rule, the appropriate assessment can only be avoided in very clear-cut cases.

The CJEU has also had the opportunity to specify the content of the assessment. In particular, all aspects of the plan or project which can affect the conservation objectives of the site must be identified in the light of the best scientific knowledge in the field.[46] Moreover, the assessment must be made considering the specific characteristics and environmental conditions of the site concerned.[47] Finally, an assessment of the implications of a plan or project for the protected site's conservation objectives is not 'appropriate', within the meaning of the first sentence of Article 6(3) of the Habitats Directive, where updated data concerning the protected habitats and species is lacking.[48] To sum it up, the assessment must be free of lacunae.[49] All of these findings are just another way of describing that all the relevant aspects of the case need to be examined carefully.

Second sentence: authorisation

The second sentence of Article 6(3) of the Habitats Directive provides that in the light of the conclusions of the assessment of the implications for the site and subject to the provisions of Article 6(4), the competent national authorities shall agree to the plan or project only after having ascertained that it will not adversely affect the integrity of the site concerned and, if appropriate, after having obtained the opinion of the general public.

Although under this provision the authorities only need to obtain the opinion of the general public 'if appropriate', it must be interpreted in the light of Article 6(1)(b) of the Aarhus Convention, providing for public participation in decisions on proposed activities which may have a significant effect on the environment. Therefore, the Court has concluded that Article 6(3) of the Habitats Directive creates a right to participate in a procedure to authorise a plan or project that could significantly affect a Natura 2000 site.[50] As this is the condition that triggers the appropriate assessment all such assessments should be conducted with public participation.

The core of the second sentence of Article 6(3) of the Habitats Directive is the substantive rule on the granting of permits. The authorities shall agree to the plan or project only after having ascertained that it will not adversely affect the integrity of the site concerned.

The first issue in this regard is the concept of the integrity of a site. The Court has found that in order for the integrity of a site as a natural habitat not to be adversely affected, the site needs to be maintained at a favourable conservation status. This entails the lasting preservation of the constitutive characteristics of the site that are connected to the presence of a natural habitat type whose preservation was the objective justifying the designation of that site in the list of sites of Community importance.[51] The same must apply, *mutatis mutandis*, to protected species.[52] Therefore, the conservation objectives of a site determine its integrity for the purpose of Article 6(3) of the Habitats Directive.[53]

Nevertheless, even with this specification, we again arrive at a point where complex assessments under conditions of uncertainty are necessary. Because our knowledge is limited, it may be possible to develop more or less substantiated hypotheses about effects, but in many cases, there will also be reasonable scientific doubts or even equally substantiated alternative hypotheses. A fish ladder that aims to mitigate the effects of a barrier restricting fish migration in a river can serve as an illustration. While there seems to be a justified expectation that such a measure, in principle, will reduce the impact of a barrier, it appears that the effectiveness of a specific ladder can only be determined by subsequent monitoring.[54] In addition, such assessments typically involve forecasts about future developments in the absence of complete information about all relevant factors. For example, the effectiveness of a fish ladder can depend on the weather conditions and the quantities of water available in the river in question. These cannot be known in advance when a decision on the construction of the fish ladder is taken. Under the general system of judicial review described above, this uncertainty should indicate wide discretion on the part of the competent authorities. Indeed, it appears that the Netherlands in the leading case argued that only obvious doubts could justify the refusal of a permit, calling to mind the standard of manifest error.[55]

The Court, however, took a different approach and focussed on the condition that the authorities need to *ascertain* or make sure that the integrity of the site would not be affected. So, where doubt remains as to the absence of adverse effects on the integrity of the site, the competent authority will have to refuse authorisation.[56] As regards the concept of doubt, the Court specifies that reasonable scientific doubt must be excluded.[57] The instrument to achieve this is, of course, the appropriate assessment, and therefore the Court highlights that it must contain complete, precise and definitive findings capable of removing all reasonable scientific doubt as to the effects of the works proposed on the protected site concerned.[58] This criterion and its interpretation by the Court integrate a strongly precautionary approach into the rules on site protection.[59] Though there is some residual scope for discretion with regard to the question of whether doubt can be qualified as reasonable this discretion is extremely limited if the jurisprudence of the Court is taken seriously. As soon as doubt is not completely hypothetical, but is a plausible reading of scientific evidence it should be considered reasonable. Courts should be able to verify whether any doubt crosses this threshold.

Studied in isolation, this reading of the second sentence of Article 6(3) of the Habitats Directive indeed appears to contradict the general jurisprudence on complex assessments under conditions of uncertainty. From this perspective, only the precise wording of Article 6(3), in particular, the term "ascertain", and the precautionary principle seem to justify this specific jurisprudence. However, as the precautionary principle should be read into all provisions of EU environmental law, it could be argued that in this area of law the discretion proclaimed by the Court should be significantly more restricted than is apparent from the jurisprudence. In fact, there are recent cases on other areas of EU environmental law that adopt this strict approach,[60] though the Court has refrained from applying it across the board.[61]

Derogation – Article 6(4) of the Habitats Directive

Nevertheless, already a thorough reading of the second sentence of Article 6(3) of the Habitats Directive demonstrates that the provision should not be understood in isolation as it explicitly states that the decision of the authorities is subject to Article 6(4). It is this latter provision that resolves the tensions that appear to result from the jurisprudence on Article 6(3).

Article 6(4) of the Habitats Directive provides that the Member State shall take all compensatory measures necessary to ensure that the overall coherence of Natura 2000 is protected, if, in spite of a negative assessment of the implications for the site and in the absence of alternative solutions, a plan or project must nevertheless be carried out for imperative reasons of overriding public interest, including those of a social or economic nature.

This provision focusses on the other area where competent authorities generally enjoy wide discretion, namely balancing. It allows permits for plans or projects that must be carried out for imperative reasons of overriding public interest, including those of a social or economic nature. The range of justifications is rather broad and explicitly includes economic and social reasons thus going beyond possible justifications under the older Birds Directive.[62] Some limitation is introduced by the qualification of permitted reasons as being in the "public interest". However, many private interests will also include public interest components, such as job creation, tax revenue or respect of private fundamental rights. It is therefore much more important that the terms "imperative" and "overriding" both indicate that the objectives of the measure must be more important than the impacts on the site. Otherwise, the reasons would not override the conservation objectives.[63]

The absence of alternative solutions is a necessary corollary to the overriding and imperative nature of the reasons because they would not be imperative if the objectives could be achieved with less impact on the conservation objectives of a site. Finally, the compensatory measures should help to guarantee that any harm to Natura 2000 as a whole will be prevented or at least kept to a minimum. It is important that a derogation is only possible if all three conditions are present. Without imperative reasons of overriding public interest or in the presence of alternative solutions, compensatory measures cannot justify a derogation, even if these measures are substantial and generous.

On paper, Article 6(4) of the Habitats Directive is an example of the mathematical nature of legal rules. It has even been associated with the mathematical concept of "no net loss" because it aims to guarantee that overall the conservation objectives of Natura 2000 are maintained.[64] The practical application, however, can rarely be reduced to a mathematical calculation.

We have already seen that the appropriate assessment of the plan or project at issue will probably not provide certainty about all the impacts on the relevant site(s), and in many instances indicate reasonable scientific doubt of different degrees. In the example of the river barrier with a fish ladder, some areas of a protected habitat type might be lost definitively in the dammed part of the river, but, in addition, there may be a probability of 80% that a protected fish species will not pass the barrier because it doesn't use the ladder and other areas of protected habitat types will deteriorate in ways that are hard to quantify.

The authorities must balance all of these impacts against the expected advantages of the plan or project. Mostly, these advantages will be of a completely different nature compared to impacts on a site and the authorities must identify them by way of a forecast or projection. If the barrier in our example is part of a hydropower project, there will be estimates of the power generation. In addition, the barrier could reduce the risk of flooding. There are no clear rules regarding how to balance these advantages against impacts on a protected site.

The balancing moves to an even higher level of complexity when the authorities consider if alternatives are absent. For each potential alternative they must, in principle, assess the impacts on Natura 2000 as well as all other advantages and disadvantages, thus multiplying the assessment of the plan or project itself. Moreover, the decision of whether an alternative exists or is absent requires another round of balancing, this time between the plan or project itself and the alternative. The authorities can only exclude an alternative if they arrive at the conclusion that on balance the plan or project itself is to be preferred. Sometimes this exercise may be easy because there is no alternative that would achieve the objectives of the plan or project to a sufficient degree, but at other times the outcome of this balancing exercise may be much less evident. Could the authorities, for example, exclude an alternative that increased costs by 10%, but reduced impacts on a site by the same margin? And how should they deal with the uncertainties that affect both estimates?

Uncertainty also affects the issue of compensation. On the one hand, the uncertain nature of the projected impacts on Natura 2000 that need to be compensated has already been analysed. On the other hand, the effectiveness of compensatory measures is subject to a similar degree of uncertainty. However, in both regards compensation is, in principle, less sensitive if the authorities can adjust the measures afterwards should monitoring demonstrate that a forecast or projection was inaccurate. Nevertheless, caution is necessary if insufficient compensatory measures could result in irreversible damage, for example by destroying a population of a protected species if compensation for their habitat proves to be unsuitable.

In addition to these issues related to uncertainty, compensation can also pose problems with regard to balancing. If impacts on conservation objectives can be compensated very easily and with a high degree of certainty, the balance will tilt in favour of the plan or project. Conversely, it should be much more difficult to justify impacts on conservation objectives that cannot reasonably be recreated elsewhere, such as limestone pavement,[65] or only over very long periods, such as bog habitats. If the affected habitat types or species are very rare and/or the effectiveness of compensatory measures is uncertain, justification will be even more difficult.

The complexity of this process can be illustrated by the example of a hydropower project. If the authorities need to decide under Article 6(4) of the Habitats Directive whether the construction of a barrier in a river for a hydroelectric power plant is justified by imperative reasons of overriding public interest, namely clean energy, they will need to balance this interest against the impact on protected sites. Obviously, a fish ladder would be a possible mitigation measure to allow access to an upstream site, but it would also introduce additional uncertainty into the assessment. Any estimate of the number of fish passing the barrier depends on the assessment of the effectiveness of the ladder. Likewise, uncertainty may affect the expected benefits of a project. In the example of the hydroelectric power plant, energy production could hinge on water levels in the river and these will depend on the weather. By its nature, there is no easy or obvious outcome to this balancing exercise. How much energy needs to be obtained to justify a certain reduction in the number of fish reaching their breeding areas upstream? And how should scientific uncertainty be taken into account? What about alternatives: Comparison with alternative hydropower projects should not be too difficult, but can other power sources, such as solar or wind power and biofuels with their own uncertainties and impacts on the environment be excluded? If other sources are being taken into account, should fossil power sources or nuclear power also be considered?

We see that the application of Article 6(4) of the Habitats Directive will rarely deliver a definitive substantive result that can be qualified as right or wrong. Most cases will be characterised by scientific forecasts or projections of positive and negative effects and by political judgements about the proper balance between competing interests. Nevertheless, somebody must decide under Article 6(4) – either in favour or against a plan or project. Moreover, the need for such decisions has increased because the strict standards of the jurisprudence on Article 6(3) have reduced the potential for permits to be granted under the latter provision.

Finally, in line with the principle of subsidiarity,[66] there is no indication in Article 6(4) that the relevant decisions should be taken by EU institutions.[67] On the contrary, the first paragraph of Article 6(4) only provides that Member States inform the Commission about the compensatory measures adopted, thus excluding Commission involvement in the actual balancing exercise. The second paragraph of Art 6(4) provides for increased Commission involvement if priority habitat types or species are present and considerations other than human health, public safety and environmental benefits are to be invoked. Nevertheless, this increased role is explicitly limited to the provision of an *opinion* from the Commission, excluding decision-making powers. Therefore, the responsibility for the application of Article 6(4) and the corresponding discretion lie with the Member State concerned and, in particular, with the competent authority chosen by the Member State. Incidentally, this consideration should also help to specify the objective and therefore the legal effect of the Commission opinion. Even if the Commission has no decision-making power, it should nevertheless contribute specific expertise. Given the nature of the Commission, this expertise primarily covers two areas. On the one hand, the Commission must provide a European perspective on Natura 2000: What is the contribution of the site concerned to the network overall and what needs to be done to guarantee that this contribution is maintained? On the other hand, the Commission has legal expertise and an overview of best practice in all Member States. Obviously, the Commission should also aim to contribute other relevant information that it has at its disposal, but it is unlikely that this information would be more relevant than the information that the competent authorities can obtain directly on site. Admittedly, the Commission can exert influence with such opinions because it has the power to initiate infringement proceedings if Member States do not comply with the opinion. However, the failure to comply with the opinion alone should not be sufficient to obtain a favourable judgment in such proceedings.[68]

It is true that the Court has underlined that, as an exception to the general rule of Article 6(3) of the Habitats Directive, Article 6(4) must be interpreted strictly.[69] In particular, the Court has found that works intended for the location or expansion of an undertaking satisfy those conditions only in exceptional circumstances[70] and that the mere construction of infrastructure designed to accommodate a management centre cannot constitute an imperative reason of overriding public interest.[71] More generally, the Court does not accept that economic costs alone are a determining factor.[72] Nevertheless, if courts were to strictly scrutinise decisions based on Article 6(4) they would definitely need to put themselves into the position of the competent authorities. Therefore, these findings of the Court should primarily be understood as reminders that the courts need to make sure that the limits of the discretion under Article 6(4) must be enforced strictly. In particular, the statement regarding the management centre reflects a very high likelihood that alternative locations for such projects exist where they would not affect Natura 2000. Such likelihood could reverse the burden of proof for a manifest error, resulting in the need to exclude such an error. Conversely, the Court has recognised that, in principle, irrigation[73] and the supply of drinking water[74] can be considered imperative reasons of overriding public interest that could be sufficient to justify severe impacts on a natural fluvial ecosystem and the associated Natura 2000 sites.[75] These findings implicitly recognise that the balancing of these considerations remains the responsibility of the Member States. Therefore, Article 6(4) of the Habitats Directive should allow for a broad margin of discretion where, in principle, EU law requires substantive review only with regard to manifest errors. Nevertheless, by way of analogy with the jurisprudence on the limitation of substantive discretion in certain cases where fundamental rights are strongly affected, it cannot be excluded that stricter scrutiny may be required if very rare species or habitat types are concerned. It appears questionable whether the classification as a priority species or habitat type would be sufficient to trigger this stricter scrutiny because some priority species or habitat types are actually relatively common.[76] Therefore, the need for stricter scrutiny would need to be established on a case-by-case basis.

Primarily, however, judicial review of decisions under Article 6(4) of the Habitats Directive should focus on procedural issues as well as the careful and impartial examination of all the relevant facts. Therefore, the Court insists that knowledge of the effects of a plan or a project in the light of the conservation objectives relating to a given site is an essential prerequisite for the application of Article 6(4), since, in the absence thereof, no condition for application of that derogating provision can be assessed. The assessment of any imperative reasons of overriding public interest and that of the existence of less harmful alternatives require weighing up against the damage caused to the site by the plan or project under consideration. In addition, in order to determine the nature of any compensatory measures, the damage to the site must be precisely identified.[77] In other words: Without an appropriate assessment of the plan or project, and of the alternatives, Article 6(4) cannot be invoked. In addition, we have already seen that the assessment comes with another procedural requirement, the consultation of the public. Of course, the second paragraph of Article 6(4) adds another consultation, namely the consultation of the Commission that needs to be undertaken in certain cases. Finally, it should be noted that necessary assessments and consultations must be reflected in the reasons given for any decision under Article 6(4). Moreover, the reasoning must describe the conclusions drawn from these steps and lay out how the competent authority strikes the necessary balance. In the end, it is quite unlikely that judgment calls, adequately explained by the competent authorities, will carry the risk of successful judicial challenge, but gaps or contradictions in the reasoning certainly will.

Conclusion

We have seen that, although Article 6(3) of the Habitats Directive provides a very restrictive framework for the authorisation of plans or projects that could affect Natura 2000 sites, Article 6(4) allows other considerations to be taken into account and provides for broad discretion in this regard. In the light of this understanding, the occasional accusation that the jurisprudence on site protection is too rigid[78] should be re-evaluated. The same applies to the reluctance to apply Article 6(4).[79] This reluctance is even more surprising because Member States introduced this provision into the Habitats Directive to counter the rigidity of the protection regime established by the Birds Directive.[80] Finally, it should be noted that a derogation under Article 6(4), based on a correctly conducted appropriate assessment according to Article 6(3), would provide significant legal certainty to a plan or project. Conversely, a permit based on a deficient appropriate assessment would be open to direct challenge or later review under Article 6(2) of the Habitats Directive.[81]

Notes

1 For two recent examples see Cases C-723/17 *Craeynest and Others v Brussels Hoofdstedelijk Gewest and Others*, EU:C:2019:168, Opinion of Advocate General Kokott, para. 41; C-197/18 *Wasserleitungsverband Nördliches Burgenland and others,* EU:C:2019: 274, Opinion of Advocate General Kokott, para. 95.
2 Council Directive 92/43/EEC on the conservation of natural habitats and of wild fauna and flora [1992] OJ L 206/7, as amended by Council Directive 2013/17/EU adapting certain directives in the field of environment, by reason of the accession of the Republic of Croatia [2013] OJ L 158/193.
3 Paul F.M. Opdam, Mirjam E.A. Broekmeyer and Fred H. Kistenkas, 'Identifying Uncertainties in Judging the Significance of Human Impacts on Natura 2000 Sites' (2009) 12(7) *Environmental Science & Policy* 915–916; Christoph Sobotta, 'The European Union Legal Boundaries for Semi-natural Habitats Management in Natura 2000 Sites' (2018) 43 *Journal for Nature Conservation* 263.
4 Art 1(8) of Directive 2009/30/EC of the European Parliament and of the Council of 23 April 2009 amending Directive 98/70/EC as regards the specification of petrol, diesel and gas-oil and introducing a mechanism to monitor and reduce greenhouse gas emissions and amending Council Directive 1999/32/EC as regards the specification of fuel used by inland waterway vessels and repealing Directive 93/12/EEC [2009] OJ L 140/88.
5 See Case C-343/09 *Afton Chemical,* EU:C:2010:419, paras 50ff.
6 See to that effect Cases 33/76 *Rewe-Zentralfinanz and Rewe-Zentral,* EU:C:1976:188, para. 5; C-93/12 *Agrokonsulting,* EU:C:2013:432, paras 35 and 36; C-572/16 *INEOS Köln,* EU:C:2018:100, para. 42.
7 Rolf Svedberg, 'Natural Sciences in Environmental Law' (2018) 27(4) *European Energy and Environmental Law Review* 140.
8 Tiina Paloniitty and Kangasmaa Sinikka, 'Securing Scientific Understanding: Expert Judges in Finnish Environmental Administrative Judicial Review' (2018) 27(4) *European Energy and Environmental Law Review* 125.
9 Franziska Grashof, 'The 'You Know Better' Dilemma of Administrative Judges in Environmental Matters–A Note on the German Legal Context' (2018) 27(4) *European Energy and Environmental Law Review* 151.
10 If Member State courts, based on such a strict standard of review, effectively question requirements flowing from EU law, they should and in some cases must refer appropriate questions on these issues to the CJEU under Article 267 TFEU.
11 See to that effect case C-55/06 *Arcor,* EU:C:2008:244, paras 164–169.
12 See with regard to the UK system Emma Lees, 'Allocation of Decision-Making Power under the Habitats Directive' (2016) 28(2) *Journal of Environmental Law* 206 and 209ff.
13 Convention on access to information, public participation in decision-making and access to justice in environmental matters of 1998 [2005] OJ L 124/4.
14 Case ACCC/C/2017/156 United Kingdom, pending. Paragraph 126 of the findings and recommendations with regard to communication ACCC/C/2008/33 concerning compliance by the United Kingdom of Great Britain and Northern Ireland, adopted on 24 September 2010, already suggested that the application of a "proportionality principle" could be more appropriate in environmental cases.

15 Findings and recommendations with regard to communication ACCC/C/2013/90 concerning compliance by the United Kingdom of Great Britain and Northern Ireland. Adopted by the Compliance Committee on 26 July 2021.
16 Cases C-384/16 P *European Union Copper Task Force v Commission,* EU:C:2018:176, para. 117; C-396/17 *Leitner,* EU:C:2019:375, paras 59 and 60.
17 Cases C-120/97 *Upjohn,* EU:C:1999:14, para. 35; C-211/03, C-299/03 and C-316/03 to C-318/03 *HLH Warenvertrieb and Orthica,* EU:C:2005:370, para. 76; and to that effect C-379/08 and C-380/08 *ERG and Others,* EU:C:2010:127, paras 60 and 61; C-544/15 *Fahimian,* EU:C:2017:255, para. 46.
18 Cases C-326/05 P *Industrias Químicas del Vallés* v *Commission,* EU:C:2007:443, para. 75; C-425/08 *Enviro Tech (Europe),* EU:C:2009:635, paras 47 and 62; *Afton* (n 5), para. 28; C-77/09 *Gowan Comércio Internacional e Serviços,* EU:C:2010:803, para. 55; C-78/16 and C-79/16 *Pesce and Others,* EU:C:2016:428, para. 49; C-44/16 P *Dyson* v *Commission,* EU:C:2017:357, para. 53.
19 Supra, n 14 and n 15.
20 Cases *Enviro Tech (Europe)* (n 18), para. 47, *Afton* (n 5), para. 28; C-15/10 *Etimine,* EU:C:2011:504, para. 60; C-73/11 P *Frucona Košice* v *Commission,* EU:C:2013:32, para. 75; C-223/17 P *Lubrizol France* v *Council,* EU:C:2018:442, para. 38.
21 Juliane Kokott and Christoph Sobotta, 'The Evolution of the Principle of Proportionality in EU Law – Towards an Anticipative Understanding?', in S. Vogenauer and S. Weatherill (eds.), *General Principles of Law* (Hart 2017) 169; J. Kokott and C. Sobotta, 'Judicial Review and Institutional Balance with Regard to European Monetary Policy', in European Central Bank (ed.), (2017) *ECB Legal Conference, Shaping a New Legal Order for Europe–A Tale of Crisis and Opportunities* 108.
22 Cases C-62/14 *Gauweiler and Others,* EU:C:2015:400, paras 68, 74, 81 and 91; C-493/17 *Weiss and Others* EU:C:2018:1000, para. 24.
23 Cases C-269/90 *Technische Universität München,* EU:C:1991:438, para. 14; *Industrias Químicas del Vallés* (n 18), paras 76 and 77; *Enviro Tech (Europe)* (n 18), para. 62.
24 Case C-350/12 P *Council* v *In't Veld,* EU:C:2014:2039, para. 63.
25 Cases C-604/12 *HN,* EU:C:2014:302, paras 49 and 50; C-141/12 *YS,* EU:C:2014:2081, paras 67 and 68.
26 Cases C-277/11 *M.,* EU:C:2012:744, para. 87; C-249/13 *Boudjlida,* EU:C:2014:2431, para. 36.
27 Cases *Technische Universität München* (n 23), para. 14; C-413/06 P *Bertelsmann and Sony Corporation of America* v *Impala,* EU:C:2008:392, para. 69; C-584/10 P, C-593/10 P and C-595/10 P *Commission and Others v Kadi,* EU:C:2013:518, para. 100; C-300/11 *ZZ,* EU:C:2013:363, para. 53.
28 *Afton* (n 5), paras 34, 36 and 37.
29 See in particular Directive 2011/92/EU of the European Parliament and of the Council of 13 December 2011 on the assessment of the effects of certain public and private projects on the environment [2012] OJ L 26/1 (the EIA Directive) and Directive 2001/42/EC of the European Parliament and of the Council of 27 June [2001] OJ L 197/30 (the SEA Directive).
30 For example, the Aarhus Convention (n 13) and the Espoo Convention on environmental impact assessment in a transboundary context of 1991 [2002] OJ C 104/7.
31 *Afton* (n 5), para. 33; C-5/16 *Poland v Parliament and Council,* EU:C:2018:483, para. 151.
32 Cases C-333/08 *Commission v France,* EU:C:2010:44, para. 92; *Afton* (n 5), para. 60; C-487/17 *Verlezza and Others,* EU:C:2019:270, para. 57.
33 Case C-293/12 and C-594/12 *Digital Rights Ireland and Others,* EU:C:2014:238, para. 47.
34 Directive 2006/24/EC of the European Parliament and of the Council of 15 March 2006 on the retention of data generated or processed in connection with the provision of publicly available electronic communications services or of public communications networks and amending Directive 2002/58/EC [2006] OJ L 105/54.
35 Case *Digital Rights Ireland and Others* (n 33), para. 48.

36 Cf. Case C-416/10 *Križan and Others,* EU:C:2013:8, paras 113–115.
37 Cf. Cases C-117/02 *Commission v Portugal,* EU:C:2004:266, para. 87; C-508/03 *Commission v United Kingdom,* EU:C:2006:287, para. 91; C-212/18 *Prato Nevoso Termo Energy,* EU:C:2019:898, para. 43.
38 Cases C-488/15 *Commission v Bulgaria (PM10),* EU:C:2017:267, para. 106, and EU:C:2016:862, Opinion of Advocate General Kokott, para. 96, *Craeynest* (n 1), para. 55.
39 But see on site management Sobotta (n 3) 265 and 266.
40 Cases C-127/02 *Waddenvereniging and Vogelbeschermingsvereniging,* EU:C:2004:482, para. 34; C-441/17 *Commission* v *Poland (Białowieża Forest District),* EU:C:2018:255, para. 108.
41 Cases C-258/11 *Sweetman and Others,* EU:C:2013:220, para. 30; C-521/12 *Briels and Others,* EU:C:2014:330, para. 20.
42 Case *Waddenvereniging* (n 40), para. 43.
43 Cf. n 3.
44 Cases *Waddenvereniging* (n 40), para. 44; C-293/17 and C-294/17 *Mobilisation for the Environment and Others,* EU:C:2018:882, para. 109.
45 Case C-323/17 *People Over Wind and Sweetman,* EU:C:2018:244, paras 35ff.
46 Cases C-387/15 and C-388/15 *Orleans and Others,* EU:C:2016:583, para. 51; C-142/16 *Commission* v *Germany (Moorburg),* EU:C:2017:301, para. 57; *Białowieża* (n 40), para. 113.
47 Cases *Waddenvereniging* (n 34), para. 49, and of 17 April 2018, *Białowieża* (n 40), para. 112.
48 Case *Białowieża* (n 40), para. 137.
49 Cases *Sweetman* (n 41), para. 44; *Orleans* (n 46), para. 50; *Białowieża* (n 40), para. 114.
50 Cases C-243/15 *Lesoochranárske zoskupenie VLK,* EU:C:2016:838, para. 49; People Over Wind (n 45), para. 39.
51 Cases *Sweetman* (n 41), para. 39; *Briels* (n 41), para. 21; *Orleans* (n 46), para. 47; *Białowieża* (n 40), para. 116.
52 Case C-293/17 and C-294/17 *Mobilisation for the Environment and Others,* EU:C:2018:622, para. 57, Opinion of AG Kokott.
53 Case C-461/17 *Holohan and Others,* EU:C:2018:883, paras 35–37.
54 Cf Case *Moorburg* (n 46), para. 37; David W. Roscoe and Scott G. Hinch, 'Effectiveness Monitoring of Fish Passage Facilities: Historical Trends, Geographic Patterns and Future Directions' (2010) 11(1) *Fish and Fisheries* 12.
55 Case C-127/02 *Waddenvereniging and Vogelbeschermingsvereniging,* EU:C:2004:60, para. 93, Opinion of AG Kokott.
56 Case *Waddenvereniging* (n 40), para. 57.
57 Case *Waddenvereniging* (n 40), para. 59.
58 Cases *Sweetman* (n 41), para. 44; *Orleans* (n 46), para. 50; *Białowieża* (n 40), para. 114.
59 Case *Waddenvereniging* (n 40), para. 58.
60 Case C-674/17 *Luonnonsuojeluyhdistys Tapiola,* EU:C:2019:851, para. 66, on species protection; as well as cases C-487/17 to C-489/17 *Verlezza and Others* EU:C:2019:270, para. 60, and *Prato Nevoso Termo Energy* (n 37), para. 58, on certain questions of waste law.
61 For a more in depth discussion of this jurisprudence see Christoph Sobotta, 'Recent Applications of the Precautionary Principle in the Jurisprudence of the CJEU – A New Yardstick in EU Environmental Decision Making?' (2021) 4 *ERA Forum* 723, in particular at 733f.
62 Cf. Case C-374/98 *Commission* v *France (Basses Corbières),* EU:C:2000:670, para. 50; Ludwig Krämer, 'The European Commission's Opinions under Article 6 (4) of the Habitats Directive' (2009) 21(1) *Journal of Environmental Law* 60f.; Andrew L.R. Jackson, *Conserving Europe's Wildlife: Law and Policy of the Natura 2000 Network of Protected Areas* (Routledge 2018) 196ff. and 271.
63 Cases C182/10 *Solvay and Others,* EU:C:2012:82, para. 75; C-43/10 *Nomarchiaki Aftodioikisi Aitoloakarnanias and Others,* EU:C:2012:560, para. 121.
64 Cf. Donald McGillivray, 'Compensatory Measures under Article 6 (4) of the Habitats Directive: No Net Loss for Natura 2000?' in Charles-Hubert Born and Others (eds.),

The Habitats Directive in its EU Environmental Law Context (Routledge 2014) 101; Hendrik Schoukens and Ann Cliquet, 'Biodiversity Offsetting and Restoration under the European Union Habitats Directive: Balancing Between No Net Loss and Deathbed Conservation?' (2016) 21(4) *Ecology and Society* 10.

65 Cf. Case *Sweetman* (n 41).

66 Cf. Lees (n 12) 214.

67 Cf. Krämer (n 62) 61.

68 But see Stefan Möckel, 'The European Ecological Network "Natura 2000" and Its Derogation Procedure to Ensure Compatibility with Competing Public Interests' (2017) 23 *Nature Conservation* 87, 100, arguing for a stronger binding effect of the opinion.

69 Cases C-239/04 *Commission* v *Portugal (Castro Verde),* EU:C:2006:66, para. 35; C-304/05 *Commission* v *Italy (Santa Caterina),* EU:C:2007:532, para. 82; *Solvay* (n 63), para. 73.

70 Case *Solvay* (n 63), para. 76.

71 Ibid., para. 78.

72 Cases C-399/14 *Grüne Liga Sachsen and Others,* EU:C:2016:10, para. 77; C-411/19 *WWF Italia Onlus and Others,* EU:C:2020:580, para. 41.

73 Case *Aitoloakarnanias* (n 63), para. 125.

74 Ibid., para. 126.

75 Ibid., paras 135–139.

76 According to the European Nature Information System (https://eunis.eea.europa.eu/index.jsp), the priority butterfly species 'Jersey tiger' (*Callimorpha (Euplagia, Panaxia) quadripunctaria*) is mentioned for 1,784 Natura 2000 sites and the priority habitat type 'Alluvial forests with *Alnus glutinosa* and *Fraxinus excelsior* (*Alno-Padion, Alnion incanae, Salicion albae*)' is mentioned for 6,422 sites.

77 Cases *Santa Caterina* (n 59), para. 83; *Solvay* (n 63), para. 74; *Aitoloakarnanias* (n 63), para. 114.

78 See for example Floor M. Fleurke, 'Innovation through Precaution: The Case of the Dutch Wadden Sea' (2008) 8 *Yearbook of European Environmental Law* 131f.

79 Cf. Lees (n 12) 201; Kerstin Sundseth and Petr Roth, 'Study on Evaluating and Improving the Article 6.3 Permit Procedure for Natura 2000 Sites' (2013), 63; but see Möckel (n 68), for the German practice.

80 On the protection regime of the Birds Directive see n 62.

81 Cases C-404/09 *Commission v Spain (Spanish Brown Bear),* EU:C:2011:768, para. 121; *Grüne Liga Sachsen and Others* (n 72), paras 33, 41 and 42; *Mobilisation for the Environment* (n 44), para. 85. See also Christoph Sobotta, 'Article 6(2) of the Habitats Directive and the Review of Permits that Have Become Final' (2022) 1 *ERA Forum* 89.

6

THE TRIGGER FOR ARTICLE 6(3) APPROPRIATE ASSESSMENT

The gateway revisited

Gregory Jones KC, BL

Introduction

Upon its adoption, the Habitats Directive[1] was immediately recognised as one of the "main vehicles through which the [EU was] endeavouring to safeguard its principles of natural heritage."[2] The Habitats Directive provides the essential legal matrix protecting Natura 2000 sites. Natura 2000[3] is a network of protected areas covering Europe's most valuable and threatened species and habitats. Stretching over 18% of the EU's land area and 9% of its marine territory, it is the largest co-ordinated collection of protected areas in the world and is the home to Europe's most valuable and threatened species and habitats.[4] The Habitats Directive is now generally regarded as the established "benchmark of European nature conservation law."[5]

New developments which might impact upon EU protected sites[6] are principally controlled by the process created by Articles 6(3) and 6(4) of the Habitats Directive.[7] For

> [w]hereas Article 6(1) and 6(2) concern the day-to-day management and conservation of Natura 2000 sites, Articles 6(3) and 6(4) lay down the permit procedure to be followed in cases where a plan or project, not directly connected with or necessary to the management of the site, is likely to have a significant effect thereon, either individually or in combination with other plans or projects. Such plans or projects shall be subject to an appropriate assessment of its implications for the site in view of the site's conservation objectives.[8]

The process of "Appropriate Assessment" ("AA") under Article 6(3) is at the heart of this mechanism by which habitats and their species are protected from new

DOI: 10.4324/9780429299100-6

development. Given that the Habitats Directive is tasked with the protection of the most vulnerable and precious elements of Europe's environment it might be thought surprising that it does not contain the same detailed prescriptions as to the procedure for and content of an AA that is found for environmental impact assessments in the EIA and SEA Directives.[9] Lord Carnwath commented on the absence of a specific format for AA in *Champion*[10] and the same observation was also made by Dove J in *Canterbury City Council*.[11]

The CJEU made clear in *Landelijke Vereniging tot Behoud van de Waddenzee v Staatssecretaris Van Landbouw, Natuurbeheer en Visserij* ("Waddenzee")[12] and subsequently that the application of AA to any particular facts should be on a precautionary basis. It has also used that principle as a guiding metwand when plugging procedural gaps. This may include sometimes "borrowing" definitions or terms from the EIA Directive, as it did in *Waddenzee* itself when it drew on the category of development which may require assessment. Indeed, the first sentence of Article 6(3) is often referred to as a "screening decision" but the word "screening" is not mentioned in the Directive. It is a phrase lifted from the EIA Directive. But it is important to note at the outset such "read across" is only permissible where it is consistent with the purpose of the Habitats Directive; not least, because it is the legal consequence of an adverse AA that makes it so very different from an environmental assessment carried out under either the EIA or SEA Directives. So whilst there is an important inter-relationship between the environmental assessment directives and the Habitats Directive, the two concepts are nonetheless distinct. This difference was noted by the Court in Case C-418/04 *Commission v Ireland*[13] when remarking that the EIA and SEA Directives

> contain provisions relating to the deliberation procedure, without binding the Member States as to the decision, and relate to only certain projects and plans....Accordingly, assessments carried out pursuant to Directive 85/337 or Directive 2001/42 cannot replace the procedure provided for in Article 6(3) and (4) of the Habitats Directive.[14]

An environmental assessment under the EIA or SEA Directives provides the decision maker with the necessary "environmental information" required by law which must be taken into account before determining a consent for a new development.[15] However, the decision itself is not determined in any way by the outcome of the environmental assessment.[16] By contrast, if the outcome of an AA is that the proposal is likely to have an adverse effect upon the integrity of the protected site, then consent for that proposal cannot be given unless the criteria in Article 6(4) are satisfied. As Carnwath JSC writing extra judicially observed:

> [This requirement] is particularly significant...because it imposes obligations not only on how the decision –making process must be carried out but also on the decision-making outcome.[17]

This is a very different concept from the traditional approach to town planning decision-making adopted in the UK and Ireland. Seen simply as a part of administrative law decision making the weight given to individual factors is a matter solely for the decision takers making almost any decision possible withhold any prohibitive environmental harm threshold.[18]

Accordingly, it is critical to the efficacy of this protective system that the trigger for AA is not avoided in circumstances incompatible with the overarching Treaty requirement of securing a high level of environmental protection in general and the almost absolute guarantee of protection for protected sites.[19] In *Landelijke Vereniging tot Behoud van de Waddenzee v Staatssecretaris Van Landbouw, Natuurbeheer en Visserij* ("Waddenzee")[20] the CJEU made clear that:

> ...the first sentence of Art 6(3) subordinates the requirement for an appropriate assessment of the implications of a plan or project to the condition that there be a probability or a risk that the latter will have significant effects on the site concerned. In the light, in particular of the precautionary principle... such a risk exists if it cannot be excluded on the basis of objective information that the plan or project will have significant effects on the site...[21]

Later, in C-258/11 *Peter Sweetman and Others v An Bord Pleanála* ("Sweetman"),[22] Advocate General Sharpston also opined that screening must adopt a precautionary approach to exclude any risk of significant effects to a protected site escaping AA. For her, screening is a *de minimis* threshold to exclude cases where there is clearly no risk of adverse effect. She called it the "should we bother to check?" test.

This chapter[23] examines the approach taken at the so-called "screening" or "trigger" stage of the AA process in the now discredited *Hart* line of authority as to whether or not avoidance or mitigation measures can be taken into account in determining whether a plan or project is likely to have significant effects on the protected site. It is part of the thesis of this chapter that a failure properly to understand the differences in the relationship between environmental assessment and AA led some of the judiciary in the UK and Ireland astray in the *Hart* line of authorities. With a focus on the UK and Irish jurisdictions, the chapter also asks whether the judgment of the CJEU in *People Over Wind*[24] has now settled the matter in all respects. Finally, it also considers whether the *Hart* approach to the Habitats Directive may represent a wider difference of common law judicial culture which post-Brexit might lead to a divergence between the UK and the EU, even where the articles of the Directive may have been preserved for the time being in UK law by being "cut and pasted"[25] into domestic law.

The European Commission's methodological guidance

In one sense it might be said to be surprising that there could ever have been any debate on the subject of whether mitigation measures should be included in the

AA "screening" process. The European Commission's *Methodological Guidance* quoted with apparent approval AG Kokott's Opinion in *Waddenzee* that:

> In principle, the possibility of avoiding or minimising adverse effects should be irrelevant as regards determining the need for an appropriate assessment. It appears doubtful that such measures could be carried out with sufficient precision in the absence of the factual basis of a specific assessment.[26]

The European Commission's guidance[27] is, of course, not legally binding on national courts but it is nonetheless regarded as providing important advice on the approach which is likely to be correct. The opinions of the Advocates General are advisory and do not bind the CJEU or indeed the national courts. Nonetheless, the opinions tend to be highly influential and are followed in the majority of cases.[28] Even so, the reverse was true here, with domestic courts coming to the view that it was *acte clair* that mitigation measures could be taken into account.

***Hart* attack**

In *R (Hart DC) v SSCLG*[29] Sullivan J (as he then was) rejected both the views expressed by the European Commission's guidance and the AG's opinion as set out immediately above. Before examining his reasoning, it is useful to look at the context in which the issue arose.

Some have suggested[30] that there is an inherent conflict in the Directive between habitat and species conservation and improvement, and the pursuit of plans and projects. There certainly can be a tension between the two. This tension has led to the creative approach (some might say "fudge") applied in England to the concept of "mitigation." Its origins lay in the particular problems posed by the Thames Basin Heaths Special Protection Area (the "Thames Basin Heaths SPA")[31] but its solution; known as the "Suitable Alternative Natural Greenspace" (or "SANG") has since been widely applied.[32]

The Thames Basin Heath SPA is an area of open heathland, scrub and woodland located across Surrey, Hampshire and Berkshire that supports important bird breeding populations. The pressure for new housing in the South East of England is great and the implications for the protected birds of all those new dog-walking homeowners and their wandering cats meet head on. The SANG mechanism was developed at a regional level to standardise the mitigation to be provided by local authorities and developers in connection with development in the vicinity of the Thames Basin Heaths SPA. A SANG is effectively an existing open space, that is due to undergo enhancements usually financed by local planning authorities and developers to attract more visitors by providing an enjoyable natural environment for recreation as an alternative to the use for recreation of an SPA.

The basic principle of the SANG is that there is a "zone of influence" of 400 m to 5 km as the crow flies, measured from the perimeter of the SPA to the primary point of access to 'the curtilage of the [proposed] dwelling.' Within the zone of influence, provision and ongoing management of SANG, paid for by way of developer contributions, should be sought by local authorities in relation to new residential development to ensure that there is no 'likely significant effect' on the SPA. Co-ordination at a *supra* borough/district level was particularly important given that some 15 borough or district councils in South East England fall within the zone of influence.

The result was that development can take place without an AA if the decision maker takes into account the provision of a SANG as a means of avoiding the likely significant environmental effects which might arise if dogs, cats, etc. were not diverted from the protected habitats to the newly created SANGs. The approach was challenged by Hart District Council as being contrary to the requirements of Article 6(3) of the Habitats Directive on the grounds that mitigation and/or avoidance measures could not be taken into account at the screening stage. Sullivan J found that mitigation measures should be taken into account in screening proposals under the Habitats Regulations and Directive:

> As a matter of common sense, anything which encourages the proponents of plans and projects to incorporate mitigation measures at the earliest possible stage in the evolution of their plan or project is surely to be encouraged. What would be the point, from the proponents' point of view, of going to the time, trouble and expense of devising specific mitigation measures designed to avoid or mitigate any effect on an SPA, and incorporating those proposals into the project, if the competent authority was then required to ignore them when considering whether an appropriate assessment was necessary?

Sullivan J reasoned that under the transposing regulations:

> ...that there is no legal requirement that a screening assessment under Regulation 48(1) must be carried out in the absence of any mitigation measures that form part of a plan or project. On the contrary, the competent authority is required to consider whether the project, as a whole, including such measures, if they are part of the project, is likely to have a significant effect on the SPA. If the competent authority does not agree with the proponent's view as to the likely efficacy of the proposed mitigation measures, or is left in some doubt as to their efficacy, then it will require an appropriate assessment because it will not have been able to exclude the risk of a significant effect on the basis of objective information.

The judgment touches on an issue which is addressed further below, namely, a question which might be not only technical but also wider in nature, namely, the extent to which mitigating aspects can be said to be actually part of the project rather than something "mitigating" the effects of the project. In any event, the judge put the matter quite firmly holding that "it would have been 'ludicrous' … to disaggregate the different elements of the package and require an appropriate assessment…" and noted, "the provisions in the Habitats Directive are intended to be an aid to effective environmental decision making, not a legal obstacle course…". In arriving at his conclusions, Sullivan J sought to draw particular strength from the Court of Appeal's ruling that it is permissible for a local planning authority or the Secretary of State to have regard to proposed mitigation measures when deciding, at the screening stage, under the EIA Directive. He referred particularly to *Gillespie v Secretary of State for Transport Local Government and the Regions & Ors*[33] where Pill LJ had stated that:

> When making his screening decision, the Secretary of State was not in my judgment obliged to shut his eyes to the remedial measures submitted as a part of the planning proposal… His decision is not… pre-determined either by the complexity of the project or by whether remedial measures are controversial though, in making the decision, the complexity of the project and of the proposed remedial measures may be important factors for consideration.[34]

In the same case Laws LJ was more cautious saying at paragraph [46] that

> …if prospective remedial measures are not plainly established and not plainly uncontroversial, then as it seems to me the case calls for an EIA.

Sullivan J also relied upon Pill LJ in *R (Catt) v Brighton and Hove City Council & Anor*[35] that:

> When forming a screening opinion, the Council were not required to ignore either the conditions proposed to limit the scope of the development or the conditions providing for ameliorative or remedial measures.

The Court also addressed the distinction between the Habitats and EIA Directives holding that:

> …Unlike an EIA, which must be in the form prescribed by the EIA Directive, and must include, for example, a non-technical summary, enabling the public to express its opinion on the environmental issues raised (see *Berkeley v the Secretary of State for the Environment* [2001] 2 AC 603 per Lord Hoffmann at page 615), an appropriate assessment under Article 6(3) and Regulation 48(1) does not have to be in any particular form (see paragraph 52 of *Waddenzee* judgment), and obtaining the opinion of the general public is optional.[36] Thus, considering proposed mitigation measures at the screening stage under Article 6(3) would not be frustrating the purpose of

> the legislation by pre-empting any particular form of inquiry, which was the particular concern expressed by Laws LJ in *Gillespie*.

The analysis based upon the absence of detailed provision in the Habitats Directive by comparison to the EIA Directive is accurate. However, the conclusion which Sullivan J draws from his analysis is flawed. As stated above, it has been this lack of detail that has led the CJEU to look at the EIA Directive for guidance, but in order to strengthen not weaken its effect. The CJEU has, of course, shown a willingness to look to the provisions of the EIA Directive to help fill gaps in the AA and to draw upon provisions in the EIA Directive in applying a purposive approach to interpreting the scope and meaning of the requirements of the Habitats Directive (see e.g. as previously mentioned in *Waddenzee*[37] on the definition of what constitutes a "plan or project"). Applying a purposive approach does not come naturally for many national judges schooled in the common law tradition[38] where the inclusion or absence of words is usually regarded as a significant factor in interpreting the scope of a particular provision. The two directives perform quite different functions. Ironically, therefore, the application of case law on EIA screening to screening for AA has in this instance led national courts to apply a less precautionary approach in the line of authorities based upon the *Hart* judgment.[39]

It is to be recalled that the need for an EIA is triggered by certain categories of development and the question is whether such developments are likely to have significant effects on the environment. It is a directive aimed at protecting the environment generally by ensuring significant effects on the environment are considered by the correct procedure. Such an approach could be said to be proportionate to the aims and purpose of the EIA Directive. By contrast, the Habitats Directive is aimed at protecting the most precious and vulnerable aspects of our environment. The occasions where a light screening will achieve the very high degree of certainty required to exclude[40] are going to be rare, in contrast to the position under the EIA Directive. The absence of prescribed detail in the Habitats Directive should not be used in order to justify what amounts to a "light touch" for screening things *out* subjected only to distant review by the courts applying *Wednesbury*.[41] The distinction between AA and EIA is a quite deliberate one. The note of the meeting of the Habitats Directive Working Party of 28–29 September 1989[42] records that:

> To take account of the concern voiced by a number of delegations … [France] proposed that, instead of [provisions applying the EIA directive to special protection areas under the Habitats Directive] …special rules for protection areas should be included in the [Habitats Directive itself] providing for an assessment of the effects which the development projects not directly connected to the management of these areas would have on the conservation potential of the areas concerned, on the understanding that

> the Member States would be making this assessment and would inform the Commission thereof….

The French Presidency followed this up on 13 October 1989 with a compromise proposal which removed all references to the EIA Directive and inserted a new article, paragraph 3 of which provided that,

> All development projects affecting or likely to affect a special protection area and which are not directly connected with or necessary to the management of the area shall be subject to detailed assessment of their impacts in relation to area conservation objectives….

At a meeting of the Working Party on 13 November, the UK successfully challenged the term "detailed assessment" replacing it with "appropriate assessment".

Reactions to *Hart*

The *Hart* approach was welcomed by the housebuilding industry and by many local planning authorities anxious to meet their housing delivery targets. It was subject to academic legal criticism.[43] However, the approach was robustly approved on several occasions by the Court of Appeal. In *Smyth v SSCLG*[44] Sales LJ referred to *Hart* as "compelling and clearly correct."[45] Indeed, any suggestion that there might even be some doubt as to its correctness was rejected:

> Mr Jones submitted that this part of the reasoning in *Hart* was wrong, or that the position under EU law was uncertain and that a reference to the CJEU should be ordered to obtain its view. I do not accept either submission. In my judgment, the reasoning of Sullivan J is compelling and is clearly correct, to the *acte clair* standard.

Richards L.J. also approved the approach in *No Adastral New Town Ltd v Suffolk Coastal DC*.[46] Indeed, the *Hart* approach to the question of "screening" under the Habitats Directive was also not doubted by the Supreme Court in *R (Champion) v North Norfolk DC*.[47] For the English courts, the *Hart* approach was a pragmatic solution which allowed things to work. The fear in areas around the Thames Basin SPA that there might be a damaging delay or restriction on the grant of planning permissions for much needed housing caused by an excessive number of AAs was thereby averted. Whether the proposed mitigation or avoidance worked to the sufficient degree required so as to obviate the need for an AA was – as Sullivan J remarked in *Hart* – a matter for the judgement for the planning authority, something with which the court itself would be slow to intervene. That too, reflects the general approach of the English courts to challenges to planning decisions. In *Suffolk Coastal*[48] the Supreme Court highlighted the "particularly unfortunate" "over-legalisation of the planning process."[49] These comments in that instance were not directed at EU environmental law-based challenges but rather disputes as to the correct meaning of English planning policy. English courts seek to discourage challenges which they consider involve the "excessive legalism infecting the planning system" (*Mansell v. Tonbridge and Malling BC*),[50] or "hypercritical scrutiny" of decisions which are "laboriously dissected in an effort to find fault."[51]

Hart travels

The *Hart* approach spread across the Irish Sea and was followed in Northern Ireland by Stephens J (as he then was) in the *Alternative A5 Alliance's Application for Judicial Review*,[52] although the decision in that case was over-turned for other breaches of the Habitats Directive. The case concerned a challenge to the consent granted for the proposed dualling of the A5 road. Having summarised Sullivan J's reasoning in *Hart*, Stephens J (as he then was) made the following point at [88]:

> ...Developments come in all forms and the approach to the screening opinion must have regard to the particular development proposed. The duty of the decision maker in the screening process is to examine the actual characteristics of the particular project. At one end of the spectrum of potential developments one may have remedial measures whose nature, availability and effectiveness are already plainly established and plainly uncontroversial and there may be circumstances in which those remedial measures can be independently enforced and monitored. At the other end of the spectrum one can have complex developments with remedial measures that are not plainly established and not plainly uncontroversial where it would be doubtful that such measures could be defined with sufficient precision in the absence of the factual basis of an appropriate assessment. There is no legal requirement that only the plainly established and plainly uncontroversial measures can be taken into account in the screening process but there comes a stage at which declining to conduct an appropriate assessment would pre-empt the very form of inquiry contemplated by the Habitats Directive and the purpose of the Directive would be frustrated.

Stephens J drew support from the application of national case law on the approach to the EIA Directive relied upon by Sullivan J, namely, to the effect that mitigation measures which were tried and trusted could be relied upon at the screening stage to exclude the need for AA. However, Stephens J also contemplated that mitigation measures which are not plainly established nor uncontroversial could be taken into account in the AA screening process. Given the need for scientific certainty in the AA process, it is difficult to see how an AA could be avoided by relying upon mitigation which is the subject of legitimate scientific dispute. Stephens J referenced the position under the EIA Directive pointing out that:

> [91]....*R (Bateman) v South Cambridgeshire District Council* [2011] EWCA 157 was a case relating to a screening exercise under the EIA Directive

but the principles equally apply to a screening exercise under the Habitats Directive. In that case Moore-Bick LJ (with whom Jackson LJ agreed) said:

> 11. … the decision taken on a screening opinion must be carefully and conscientiously considered and must be based on information which is both sufficient and accurate. The opinion need not be elaborate, but must demonstrate that the issues have been understood and considered …
>
> 20. … I think it important to bear in mind the nature of what is involved in giving a screening opinion. It is not intended to involve a detailed assessment of factors relevant to the grant of planning permission; that comes later and will ordinarily include an assessment of environmental factors, among others. Nor does it involve a full assessment of any identifiable environmental effects. It involves only a decision, almost inevitably on the basis of less than complete information, whether an EIA needs to be undertaken at all. I think it important, therefore, that the court should not impose too high a burden on planning authorities in relation to what is no more than a procedure intended to identify the relatively small number of cases in which the development is likely to have significant effects on the environment, hence the term screening opinion.
>
> 21. Having said that, it is clear from *Mellor* that when adopting a screening opinion the planning authority must provide sufficient information to enable anyone interested in the decision to see that proper consideration has been given to the possible environmental effects of the development and to understand the reasons for the decision. Such information may be contained in the screening opinion itself or in separate reasons, if necessary combined with additional material provided on request.

The passage from Moore-Bick LJ's judgment cited by Stephens J refers to the need to avoid heavy burdens being placed on authorities "in relation to what is no more than a procedure intended to identify the relatively small number of cases in which the development is likely to have significant effects on the environment, hence the term screening opinion." That may or may not be true of screening opinions under the EIA *regime*, it is not so under the Habitats Directive. Indeed, if the term "screening" is interpreted by national courts as indicating a light touch then it is better that the word "screening" is not used for Stage 1 consideration under the Habitats Directive. After all, it is a phrase which, of course, does not even appear in the Habitats Directive. Stephens J fell into the same trap as Sullivan J. He pointed out that "A screening opinion is different from an appropriate assessment which involves detailed consideration [and that the] screening opinion does not require all considerations to be mentioned."[53] But that lack of detail does not justify a lighter touch which is most unlikely to provide the sufficiency of certainty required by the Habitats Directive. Moreover, Stephens

J also took a traditional British common law approach to how the courts should treat such decisions holding that:

> [90] The decision as to whether the plan or project is likely to have a significant effect on the integrity of the specific sites or whether there is some doubt as to the efficacy of the remedial measures, is for the Minister subject to judicial review on traditional *Wednesbury* grounds.

This regards the trigger issue as very much one of planning judgement.[54] The traditional approach adopted by the courts in the UK and also in Ireland[55] to matters of planning judgement is non-interventionist. But is this the correct approach for AA screening? It would mean the rigorous requirements of AA can be avoided entirely at the preliminary stage with very little judicial supervision. Should the gateway or trigger to the AA process instead be regarded more a matter of so-called "precedent fact" and therefore be subject to objective determination by the court? Fordham[56] says of precedent fact that:

> "...the logic of precedent fact questions is of an objective factual question whose existence is needed to 'trigger' the public body's proper function. The 'trigger' concept harks back to "jurisdictional error."

Whether a person is or is not a child, has been held to be a question of precedent fact to be determined by the court.[57] However, in *R(Jones) v Mansfield*[58] the Court of Appeal in England and Wales rejected the argument that the "likely significant environmental effects on the environment" for the purposes of triggering an EIA was a precedent fact. Whilst not addressed expressly as a question of precedent fact in *Rossmore and Killross v An Bord Pleanála, the State and Eirgrid*[59] the High Court in Ireland had to consider, among other things, whether mitigation should be considered at the AA screening stage for a proposed project and if so what the standard of scrutiny should be.[60] In that case, the applicants raised four questions for certification. Hedigan J found each of the questions posed had already been clearly decided and accordingly rejected the application for certification for all four questions and refused leave to appeal. The first question was, however, most interesting:

> By what standard must the courts assess the decisions taken by a Competent Authority on an appropriate assessment and the screening for an appropriate assessment?

In bringing the challenge the applicants relied upon the following statement by Finlay Geoghegan J in *Kelly v. An Bord Pleanála*[61]:

> ... it appears to me that whilst the requirement for an appropriate assessment has been implemented in Ireland by amendment of the Planning Acts

> and requires to be carried out inter alia as part of the planning process, the determination which must be made by the Board as competent authority it is [sic] not a 'planning decision' in the sense used in the judgments relating to reasons relied upon by the Board. In such a planning decision, the Board is exercising a jurisdiction with a very wide discretion. By contrast, the determination it must make as part of an appropriate assessment is significantly narrower and legally constrained as explained in the CJEU cases cited. It also determines the Board's continuing jurisdiction to grant planning consent, and therefore a decision which goes to its jurisdiction.

Mr. Justice Hedigan's approach to this question was described as "old school."[62] he held that the court should not interfere with the Board's decision once it was satisfied that the Board had evidence before it on which it could reasonably base its decision. He said:

> It seems to me that the decision of Finlay Geoghegan J. deals with the nature of a Stage Two appropriate assessment. Such an assessment, as identified by her, requires a very specific type of analysis, evaluation and decision. This is because it occurs where, on a Stage One screening, it has been determined that it is likely a certain project will have a significant effect on a European site. Thus, she finds a stage two assessment must be accompanied by a more detailed reasoning than would occur in the wider jurisdiction of a normal "planning decision". I think the rationale of this decision is limited to stage two assessments because the stage two appropriate assessment with which it deals must analyse the probable (likely) effects the better to deal with them or justify outruling them. Where the finding is that there is no such likelihood of effects, then there is, by definition, very much less to find, analyse and assess.

Mr. Justice Hedigan was correct in identifying that Finlay Geoghegan J. was specifically addressing the Stage 2 part of the AA but the conclusions he draws from that distinction are entirely unsound. Once again, it is the absence of a specific detailed process by which the screening is to be carried out which is seized upon, in this instance to suggest that the competent authority should be given a greater margin of latitude by the courts to accept a light touch screening for AA screening purposes. Mr. Justice Hedigan does not address Finlay Geoghegan J.'s view that it is not a planning judgement in the traditional sense which is being carried out: "It [the AA] also determines the Board's continuing jurisdiction to grant planning consent, and therefore a decision which goes to its jurisdiction." True it is she is making these remarks at a later stage of the process, but the trigger to the AA process must surely also be a question of jurisdiction? The third question raised by the applicants in *Rossmore and Killross* was:

> To what extent is the Competent Authority entitled to take account of mitigation measures in the Stage One screening decision in determining that there would be no likely significant effect on an SAC?

Mr. Justice Hedigan concluded that the answer to this question was also already clearly decided:

> The first thing to be noted in regard to this question is that this Court has found as a matter of fact that the mitigation measures proposed were an intrinsic part of the work to be carried out. This being so, the issue was decided by this Court in relying upon the judgment in *Hart* [2008] EWHC 1204 (Admin.) at para. 50 onwards and notably at para. 61 that, where mitigation measures are an intrinsic part of the project in question, they may be taken into account in the stage one screening process. This decision was considered and followed in the High Court of Northern Ireland in *Alternative A5 Alliances Application for Judicial Review* [2013] NIQB 30, and, of course, in this case. Thus, there are decisions of three jurisdictions supporting this principle. There is, thus, both no uncertainty in relation to this point, nor is there any pressing need within a sparing jurisdiction to further litigate the point, notably where the work involved has already been done. Indeed, in a certain sense, the issue is now moot. In any event, there is no uncertainty or any pressing need to further litigate this point and it, therefore, also fails the test for certification.

It is to be observed that Hedigan J. grounded this aspect of his decision upon the fact that the mitigation measures were an intrinsic part of the work to be carried out. In finding that there was certainty in the matter the judge relied upon *Hart* and *Alternative A5 Alliances Application for Judicial Review*. Reference to the decisions of national courts of other Member States on points of EU law is to be lauded and encouraged. At the time of the judgment *Hart* had not been approved in the Court of Appeal in England and Wales and was a decision at first instance. Such decisions are not binding even on other English High Court judges and obviously, in any event not binding upon the Irish High Court. Mr. Justice Hedigan would have been aware of the status of *Hart*. The *Alternative A5 Alliances Application for Judicial Review* was also a judgment at High Court level in Northern Ireland. In that case, the applicants were successful in quashing the various orders on other Habitats Directive grounds. Accordingly, the applicants had no ability (or particular desire) to take proceedings to the Court of Appeal to challenge the finding of the court that screening under the Habitats Directive could take mitigation into account. Again, this would have been fully appreciated by Hedigan J. That does not mean, of course, that such first instance decisions should not be prayed in aid by a judge in another jurisdiction where the judge considers the reasoning of his/her EU colleague to be sound. However, the status of such judgments is relevant when assessing whether they can be said to be relied upon to have provided the definitive answers to a point of EU law, particularly in this instance where persuasive sources at an EU level suggesting an alternative view were apparently ignored by the judge. The *Waddenzee* judgment is quoted elsewhere in Hedigan J's ruling but there is no mention of AG Kokott's Opinion and its apparent endorsement by the European Commission's Methodological

Guidance. These authoritative, albeit non-binding statements, should have merited sufficient doubt to have warranted certification for a full hearing. The robust approach of Hedigan J nonetheless echoes the equally trenchant approach adopted soon later by the Court of Appeal of England and Wales in *Smyth*[63] declaring the matter to be *acte clair* when refusing a reference to the CJEU.

The preliminary reference in People Over Wind v Coillte Teoranta

This case concerned a challenge to cable connection to a permitted wind farm from the electricity grid. The main issue was whether river pollutants resulting from the laying of the cable would have a harmful effect on the designated River Nore pearl mussel. The planning application was screened out on the basis of expert advice. "This conclusion was reached on the basis of the distance between the proposed Cullenagh grid connection and the European sites [i.e. Natura 2000 sites], and the protective measures that have been built into the works design of the project."

When making the preliminary reference[64] in *People Over Wind v Coillte Teoranta*[65] Barrett J gave what in effect amounted to a preliminary view on the substantive issues and the reasons why the approach adopted in *Hart* troubled him into making the reference.[66] He made four key observations. First, he stated that the reasoning of Sullivan J. would appear to sit uneasily with the Opinion of AG Kokott in *Waddenzee*, para. 71. This might be thought to have been manifestly a statement of the obvious but apparently not elsewhere. Secondly, the approach taken in *Hart* relied on the decision of the Court of Appeal in England and Wales in *R. (Catt) v. Brighton & Hove City Council*.[67] Barrett J noted that the decision in *Catt* related to screening for EIA; the Habitats Directive takes a more precautionary approach than that adopted in the EIA context. Barrett J felt that a question existed as to whether it was entirely proper to import EIA-related case-law directly into the rather different context of determining whether mitigation measures should be counted in screening for appropriate assessment done within the penumbra of the Habitats Directive. Thirdly, the Judge considered that the approach adopted in *Catt* could "perhaps be criticised as being, in any event, overly subjective in determining the significance of screening." Finally, he thought the approach taken in *Hart* could ultimately limit the role of public participation in a manner not consistent with the Habitats Directive if mitigation measures were to be incorporated into project design so as to circumvent the need for full appropriate assessment and thus avoid the obligation for public notification. In my view all four points raised by Barrett J were sound. Consequently, he referred the following question to the CJEU:

> Whether, or in what circumstances, mitigation measures can be considered when carrying out screening for appropriate assessment under Article 6(3) of the Habitats Directive?

CJEU: C-323/17 People Over Wind v Coillte Teoranta

The CJEU answered the reference question as follows:

> [The Habitats Directive]...must be interpreted as meaning that, in order to determine whether it is necessary to carry out, subsequently, an appropriate assessment of the implications, for a site concerned, of a plan or project, it is not appropriate, at the screening stage, to take account of the measures intended to avoid or reduce the harmful effects of the plan or project on that site.

The Court determined that the approach adopted had been in breach of Article 6(3) since the proposals should not have been screened by taking into account the mitigation measures built into the design. The fact that these elements are described as having been "built into the design" is important because it is often said that built-in measures are not mitigation measures and can be used to screen out AA. The CJEU found:

> 25. ... Article 6 ... divides measures into three categories, namely conservation measures, preventive measures and compensatory measures, provided for in Article 6(1), (2) and (4) respectively. It is clear from the wording of Article 6 of the Habitats Directive that that provision contains no reference to any concept of 'mitigating measure'.

In so doing, the CJEU made the first obvious point that

> the fact that ... measures intended to avoid or reduce the harmful effects of a plan or project on the site concerned are taken into consideration when determining whether it is necessary to carry out an appropriate assessment presupposes that it is likely that the site is affected significantly and that, consequently, such an assessment should be carried out.[68]

That being so "a full and precise analysis of the measures capable of avoiding or reducing any significant effects on the site concerned must be carried out not at the screening stage, but specifically at the stage of the appropriate assessment."[69]

The Court also pointed to its own case-law that emphasises the fact that the assessment carried out under Article 6(3) of the Habitats Directive may not have *lacunae* and must contain complete, precise and definitive findings and conclusions capable of removing all reasonable scientific doubt as to the effects of the proposed works on the protected site concerned.[70] If the competent authority took account of mitigation measures at the screening stage this would be liable to compromise the practical effect of the Habitats Directive in general, and the assessment stage in particular, as the latter stage would be deprived of its purpose and there would be a risk of circumvention of that stage, which constitutes, however, an essential safeguard provided for by the directive.[71] Interestingly, like Barrett J the CJEU placed weight on the possibility of public participation in the process: a member of the public may derive "a right to participate in a procedure for the adoption of a decision relating to an application for authorisation of a plan or project likely to have a significant effect on the environment."[72]

Acte Clair?

Given the finding by the Court of Appeal of England and Wales that the matter was *acte clair* it is noteworthy that the CJEU took a rather different approach from the very outset. As of 2003, Advocates General are only required to give an opinion if the Court considers the case raises a new point of law.[73] It is therefore to be noted that the CJEU in *People Over Wind* clearly considered quite early in the process that the case did not raise a new point of law. In that sense, it agreed with Sales LJ and his experienced colleagues[74] in *Smyth* in the Court of Appeal and with Hedigan J in *Rossmore and Killross* in the Irish High Court, that the law was clear, but with an opposite view as to what it was clear about! No doubt in arriving at its view that it did not require the services of an Advocate General, the CJEU would have had in mind AG Kokott's Opinion on the matter and the Opinion's inclusion in the European Commission's Methodological Guidance. In addition, it would have recalled its own judgment in *Orleans v. Vlaams Gewest* C-387/15.[75] *Orleans* concerned the Antwerp Regional Development Implementation Plan (RDIP) which included works that would involve the destruction of habitats within an SAC. The Flemish Government argued that the Plan's rules meant the development of affected areas would become possible only after the sustainable establishment of habitats and habitats of species in "ecological core areas" had been confirmed as being successful and that the ecological core areas would already contribute to the integrity of the site in question. The use of those ecological core areas in the RDIP was not a compensatory measure, but rather a conservation measure. This argument was rejected by the CJEU in terms that predicted its approach in *People Over Wind*. The CJEU said:

> 48 ... it must, in the first place, be recalled that, in [29] of the judgment ... in *Briels and Others* (C-521/12), the Court held that protective measures provided for in a project which are aimed at compensating for the negative effects of the project on a Natura 2000 site cannot be taken into account in the assessment of the implications of the project provided for in art.6(3).

The CJEU was also quick to draw a clear distinction between mitigating measures which in reality amounted to compensation.

> 57 ... as noted in [33] above, that the wording of art.6 of the Habitats Directive contains no reference to any concept of "mitigating measure."

> 58 In this connection, as the Court has already observed, the effectiveness of the protective measures provided for in art. of Directive 92/43 is intended to avoid a situation where competent national authorities allow so-called "mitigating" measures — which are in reality compensatory measures — in order to circumvent the specific procedures provided for in art.6(3) and authorise projects which adversely affect the integrity of the site concerned...
>
> 59 It follows that the negative implications of a plan or project not directly connected with or necessary to the management of a special area of conservation and affecting its integrity do not fall within the scope of art.6(3) of the Habitats Directive.

Essential to the *rationale* of these cases is a refusal by the CJEU to accept that mitigation measures can be certain in their outcome.

No Straw in the Wind

People Over Wind was quickly followed by a series of judgments on AA. *Grace and Sweetman v An Bord Pleanála*[76] looked at the extent to which mitigation can be taken into account at Stage 2 of the AA. The CJEU repeated orthodox principles relating to the need for certainty in the AA. It rejected the dynamic management of the habitat which was proposed in that particular case as appropriate for passing AA and considered that approach to be compensation under Art 6(4).[77] In *Coöperatie Mobilisation for the Environment and Vereniging Leefmilieu*[78] the CJEU took the opportunity to reinforce the high levels of certainty required as regards the efficacy of mitigation before it could be taken into account at Stage 2. The CJEU makes clear that:

> 126 Moreover, according to the Court's case-law, it is only when it is sufficiently certain that a measure will make an effective contribution to avoiding harm to the integrity of the site concerned, by guaranteeing beyond all reasonable doubt that the plan or project at issue will not adversely affect the integrity of that site, that such a measure may be taken into consideration in the 'appropriate assessment'...
>
> 130 The appropriate assessment of the implications of a plan or project for the sites concerned is not to take into account the future benefits of such 'measures' if those benefits are uncertain, inter alia because the procedures needed to accomplish them have not yet been carried out or because the level of scientific knowledge does not allow them to be identified or quantified with certainty.

In *Holohan v An Bord Pleanála*[79] the CJEU went even further than before in terms of the scope of the AA, which may have to extend beyond designated habitats and the species for which the habitat has been listed.

> 40 … an 'appropriate assessment' must, on the one hand, catalogue the entirety of habitat types and species for which a site is protected, and, on the other, identify and examine both the implications of the proposed project for the species present on that site, and for which that site has not been listed, and the implications for habitat types and species to be found outside the boundaries of that site, provided that those implications are liable to affect the conservation objectives of the site.

These judgments post *People Over Wind* apply of course to Stage 2 of the AA. However, they make it even more unlikely that the very low threshold of triggering an AA can be avoided.

In *Eoin Kelly v An Bord Pleanála*[80] Barniville J gave a helpful and authoritative summary of the principles applicable to the screening stage derived from CJEU and Irish caselaw.[81] The judgment also clarified that when considering a screening report, it is the substance of the report that matters and not the use or non-use of particular words or phrases.[82] *Eoin Kelly* is also notable in providing some guidance as to whether a measure amounts to a mitigation measure or not. The use of a SUDS drainage system did not amount to a mitigation measure that had been taken into account improperly at the screening stage because

> the policy behind requiring SUDS and the inclusion of SUDS measures in a development is not in any way directed to the protection of any European site which might potentially be affected by a particular development off that site. It is clear that the key driver for the requirement to incorporate SUDS in development is the Water Framework Directive and not the Habitats Directive.[83]

Two Irish cases have subsequently cited *Eoin Kelly* and *People Over Wind* when quashing decisions on the ground that the screening assessment unlawfully took mitigation measures into account.[84] *Uí Mhuirnín v Minister for Housing Planning and Local Government*[85] is particularly notable, as after quashing the decision because the Respondent had unlawfully taken mitigation measures into account contrary to *People Over Wind*, Quinn J went on to say that,

> This result is regrettable because it appears from the material exhibited that many, if not most, of the effects of the Test Site on the environment generally will be, minimal or negligible and in some cases temporary. However, this Court must consider whether the Respondent has complied with the Habitats Directive. In this context, the critical and defining conclusion required for Article 6(3) of the Habitats Directive that there are not likely to be any significant effects on the conservation objectives of the identified Natura sites, is in the Report of the Committee expressly stated to have been made on the basis of what has come before that conclusion, which includes the mitigation measures.[86]

The Irish courts have therefore applied *People Over Wind* relatively strictly, ensuring that the threshold for triggering an AA remains low as intended by the CJEU.

The reaction in England

In England, the *People Over Wind* judgment was described variously as "controversial,"[87] "a profound shift"[88] and "a departure from previous case law particularly in the UK, with the potential to create significant issues for developers and competent authorities."[89] The National Planning Policy Framework ("NPPF") which set the government's planning policy for England was changed as a result.[90] The judgment impacts decisions to grant planning permission as well as the ongoing preparation of development plans. The implications of *People Over Wind* were examined by the English High Court for the first time in the two joined cases, *Crondall Parish Council v Secretary of State for Housing, Communities and Local Government, Crondall Developments Limited v Hart District Council,*[91] where Mr Justice Dove applied the approach of the CJEU set out in *People Over Wind* rather than that of the Court of Appeal in *Smyth* (endorsing the *Hart* approach). Interestingly the judgment suggests the same principle may well apply to screening decisions about EIA. The judge emphasised that the exercise of *Champion* discretion not to quash for justiciable errors did not permit courts to make their own planning judgements or strike new planning balances. The Secretary of State nonetheless invited the Court to exercise its discretion to not quash the planning permissions in both cases (relying on *Walton v Scottish Ministers*[92] and *Champion*[93]) on the basis that it should conclude on the facts that the decisions would not have been any different had the error not been made. When considering the discretion to not quash, the judge held that it was essential he should take "a careful and fact-sensitive examination of the available evidence." On the evidence in the *Crondall Parish Council* case, there were two key reasons why the judge refused to exercise his discretion not to quash. First, the Court was effectively being invited to re-strike the planning balance reached by the inspector which would involve the making of a planning judgement which was not a task which courts should seek to undertake. Second, the judge accepted at the Parish Council's submission that there were "loose threads" in the analysis of potential significant effects on the SPA, preventing him from concluding with certainty that the decision would have been the same without the error of law which occurred. By contrast, the Judge did exercise his discretion not to quash in the *Canterbury City Council* case because he was satisfied in that case that the error did not make a difference to the outcome.

An almost immediate head-on assault on *People Over Wind* was made by a developer and rejected by Dove J in *Gladman Developments Ltd v Secretary of State for Housing, Communities and Local Government & and Medway Council.*[94] The developers, Gladman Developments, no doubt felt particularly aggrieved by the timing of the *People Over Wind* judgment. The local planning authority, Medway

Council, had refused to grant planning permission for 225 homes at Cliffe Woods on grounds of accessibility and the adverse impact of the proposals on local character and amenity. Gladman appealed and the inspector recommended that planning permission should be granted, but a few weeks later the CJEU handed down judgment in *People Over Wind.* Following lengthy exchanges on the implications of this decision, the Secretary of State eventually refused Gladman's application. Dove J noted: "This was a departure from the domestic jurisprudence on this issue which had...held that it was permissible to take account of mitigation measures." Gladman called for a reference to the CJEU to clarify the position as it claimed that *People Over Wind* was wrongly decided, that it conflicted with domestic authorities and was inadequately reasoned and explained. The Judge however said: "I am wholly unpersuaded firstly, that there is any justification for the reference of the question proposed to the CJEU or, secondly, that in substance *People Over Wind* was wrongly decided in any event." However, there are other more subtle challenges to *People Over Wind.* Following the judgment The Planning Inspectorate, which supplies the planning inspectors who hear all and determine nearly all of the planning appeals for England and Wales and carries out independent examination of proposed local development plans, issued a practice note[95] stating:

> 11. It should be noted that there is no authoritative definition of what constitutes an integrated or additional avoidance or reduction measure and this should be considered on a case by case basis. If a measure is being introduced to avoid or reduce an effect on a European site then it can be viewed as mitigation. It may be helpful to consider whether a proposal could be considered integral to a plan or whether it is a measure to avoid harm....

Indeed, in a challenge to a licence for the culling of badgers Sir Ross Cranston (sitting as Deputy High Court Judge) in *R(Langton) v Secretary of State for the Environment, Food and Rural Affairs and Natural England*[96] rejected the contention that the licence conditions relating to the times and places for badger culling infringed the approach in *People Over Wind*:

> 157 In my view the licence conditions which Natural England attached to the licences in Areas 16 and 17 are not the mitigating or protective measures which featured in the People Over Wind ruling. They are properly characterised as integral features of the project which Natural England needed to assess under the Habitats Regulations. I accept Natural England's submission that it would be contrary to common sense for Natural England to have to assume that culling was going to take place at times and places where the applicants did not propose to do so.

Permission to appeal was granted and the case went to the Court of Appeal.[97] By the time of the hearing an AA had been carried out and the Court of Appeal refused to rule on this point on the grounds that it had become an academic point.

More recently, the necessity of carrying out an AA has been considered in *R (Hudson) v Windsor and Maidenhead RBC.*[98] In *Hudson,* the Claimant challenged the local planning authority's decision to grant permission for the construction of a holiday village at the Legoland Windsor Resort. There were a number of grounds, and in the High Court it was common ground that an AA should have been undertaken and that the local planning authority had acted in breach of the Habitats Directive by not carrying out an AA. Nonetheless, in the High Court Lang J withheld relief because under s31 (2A) of the Senior Courts Act 1981, it was highly likely that the outcome would not have been substantially different if an AA had been undertaken.[99] The application of judicial discretion in this case bears some resemblance to Dove J's refusal to quash the decision in *Canterbury City Council.* Lang J reached this conclusion in *Hudson* because the developer had carried out "extensive, detailed assessments, first identifying an adverse impact on the integrity of the SAC, and then assessing the effectiveness of the proposed mitigation measures" and because Natural England was eventually satisfied, after the submission of further material, that "the adverse effect on the integrity of the SAC could be mitigated, so as to make the development acceptable."[100] In effect, the High Court therefore decided that an AA was not necessary because mitigation measures would ensure that there was no adverse effect on the integrity of the protected site. Although this approach arguably amounts to taking mitigation measures into account when screening out an AA, the crucial passage in *People Over Wind* concerning the consideration of mitigation measures at the screening stage was not cited in the judgment.

On appeal, the Court of Appeal upheld Lang J's approach. In response to an additional point that it was not clear how and from where the buffer zone between the proposed development and the protected site had been measured, Coulson LJ held that "I am not persuaded that the accuracy of the measurements has anything to do with the absence of an AA, or the possible outcome of the planning application if there had been an AA."[101] This quotation raises questions as to how the requisite level of certainty required under the Habitats Directive could have been achieved in the absence of an AA. Furthermore, *People Over Wind* was not cited in Coulson LJ's judgment, although the court's use of s31(2A) appears to be a way for the court to take mitigation measures into account when deciding whether an AA is necessary or not. Section 31(2A) was also invoked in *R (Wingfield) v Canterbury City Council*[102] where Lang J held that even if conducting an AA at the reserved matters stage rather than the outline permission stage had been unlawful, relief would have been refused on the ground that the result of conducting an AA at the outline permission stage would have inevitably been the same. Although the use of s31(2A) was hypothetical in *Wingfield* it serves as another example of judicial discretion being invoked when considering screening assessments under the Habitats Directive.

Hudson, Canterbury City Council and *Wingfield* are all early indications of the English courts potentially departing from the high standards imposed by the CJEU by invoking judicial discretion. It certainly stands in contrast to the approach adopted by the Irish High Court in *Uí Mhuirnín,* where the court proceeded to quash the decision for being contrary to *People Over Wind* even though it described this result as "regrettable" due to the minimal environmental effects of the proposed development.

Conclusion

Lawyers everywhere like to interpret words carefully. That is not only the case for common law lawyers. Detailed provisions make it easier to ensure the uniform application of EU law. If one is having to look to grander purposive approaches in order to interpret provisions consistently at national level this is less likely to be achieved until there has been ruling from the CJEU. The absence of precision in the Habitats Directive generally, and in terms of the Article 6(3) screening process, in particular, has had the effect of encouraging judges to allow a lighter touch to be applied by competent authorities, allowing them to permit developments to proceed without AA on the basis of mitigation measures whose efficacy is supported by slight or untested evidence. Ironically, borrowing from approaches adopted by the national courts regarding screening under the EIA Directive served only to weaken the approach to the Habitats Directive at national level.

The scope of what is said to be part of the project rather than a form of mitigation is likely to be an area regarded by the English courts and perhaps also by the Irish too, as a matter of fact and degree into which they will not readily intervene. Post-Brexit, however, the Irish courts will still be subject to the supremacy of the CJEU.[103] It may be that the screening decision is not properly regarded as a so-called "precedent fact" since whether something is likely to have significant environmental effects will often involve some degree of expert judgement. However, the extraordinarily high level of certainty that is required in order to exclude the need for AA means that there is very little scope for the national courts exercising deference to the judgement or expertise of the competent authority when examining whether it was correct to screen out an AA. Further, and notwithstanding Brexit, decisions of the Aarhus Convention Compliance Committee may have an important role to play in the decisions of national courts in countries party to the Convention, including the UK and Ireland. For example, in a recent decision against the UK,[104] the Compliance Committee held that,

> 120. [...] in a challenge to the substantive legality of an EIA screening decision, the court must make its own assessment, based on all the evidence put before it, as to whether the proposed activity was likely to have a significant effect on the environment and thus to require an EIA. It would not be sufficient to merely check that the decision-maker carried out the correct procedural steps for determining whether a project was likely to have significant effects. Nor does it suffice for the court to check that the

> decision-maker had formally applied the correct legal test and that the decision-maker had convinced itself that that test was met in a particular case.
>
> 121. To be clear, the Convention does not require the court to undertake a completely fresh analysis of all matters arising in the case and to substitute its decision for the decision taken by the competent authority. Nevertheless, the court must undertake its own assessment of all the evidence before it to determine whether the applicable legal requirements were met. The Committee considers that this requires the court to perform a review function over findings of fact and the weight to be given to evidence where those may have a direct impact on the determination as to whether the applicable legal test (for example, likely significant effects) has been met.

This interpretation of obligations pursuant to the Aarhus Convention would certainly appear to challenge the traditional deference shown to expert decision-makers by the courts in the UK and Ireland, and may require a change of approach by those courts. Thus, while the UK courts may no longer be subject to the supremacy of the CJEU, they are still subject to the influence of instruments of international law to which the UK remains party.

How environmental legislation develops in the UK post-Brexit is presently impossible to predict, albeit the Government recently revealed proposals to move away from the EU law approach to appropriate assessment.[105] Equally interesting, however, will be the approach of the UK courts to the interpretation of existing EU environmental law within UK law, now that the UK courts have been removed from the supremacy of the CJEU. The *Hart* saga suggests that the UK courts are most likely to begin divergence in areas where the legislation is lacking in detailed wording. The examples of *Hudson, Canterbury City Council,* and *Wingfield* also suggest that the UK courts may have already begun to depart from the strict standards set by the CJEU.

Notes

1 Council Directive 92/43/EEC of 21 May 1992 on the conservation of natural habitats and of wild fauna and flora.

2 Jonathan Faulks, "The EU Habitats Directive" European Environment, Spring 1994, 12–26.

3 The sites in the Natura 2000 network are designated under the "Nature Directives," i.e. the Birds and the Habitats Directives. In 1979, the Birds Directive (amended in 2009) established an EU-wide protection regime for all bird species naturally occurring in the EU including classification by Member States of Special Protection Areas (SPA). This approach was extended through the 1992 Habitats Directive, which also provided for the establishment of a representative system of legally protected areas throughout the EU. These areas are named Sites of Community Importance (SCI) and aim for the conservation of the 233 habitat types listed in Annex I of the Directive and the 900 plus species listed in Annex II. SCIs must also be designated as Special Areas of Conservation (SAC) as soon as possible and within six years at most. SPAs and SCIs/SACs together make up the Natura 2000 network. Post Brexit,

changes have been made to the Conservation of Habitats and Species Regulations 2017 (as amended) by the Conservation of Habitats and Species (Amendment) (EU Exit) Regulations 2019. The main changes to the 2017 Regulations are: the creation of a national site network within the UK territory comprising the protected sites already designated under the Nature Directives, and any further sites designated under these Regulations; the establishment of management objectives for the national site network (the 'network objectives'); a duty for appropriate authorities to manage and where necessary adapt the national site network as a whole to achieve the network objectives; an amended process for the designation of SACs; arrangements for reporting on the implementation of the Regulations; given that the UK no longer provides reports to the European Commission, arrangements replacing the European Commission's functions with regard to the imperative reasons of overriding public interest (IROPI) test where a plan or project affects a priority habitat or species; arrangements for amending the schedules to the Regulations and the annexes to the Nature Directives that apply to the UK.

4 These figures apply to the EU of 27 countries excluding the UK.

5 Emma Lees, "Allocation of Decision-Making Power under the Habitats Directive" (2016) 28 *Journal of Environmental Law* 191–219.

6 Special Areas of Conservation and Special Protection Areas.

7 3. Any plan or project not directly connected with or necessary to the management of the site but likely to have a significant effect thereon, either individually or in combination with other plans or projects, shall be subject to appropriate assessment of its implications for the site in view of the site's conservation objectives. In the light of the conclusions of the assessment of the implications for the site and subject to the provisions of paragraph 4, the competent national authorities shall agree to the plan or project only after having ascertained that it will not adversely affect the integrity of the site concerned and, if appropriate, after having obtained the opinion of the general public.

4. If, in spite of a negative assessment of the implications for the site and in the absence of alternative solutions, a plan or project must nevertheless be carried out for imperative reasons of overriding public interest, including those of a social or economic nature, the Member State shall take all compensatory measures necessary to ensure that the overall coherence of Natura 2000 is protected. It shall inform the Commission of the compensatory measures adopted."

8 European Commission, "Article 6 of the Habitats Directive Rulings of the European Court of Justice" (Guidance), FINAL DRAFT September 2014. https://ec.europa.eu/environment/nature/info/pubs/docs/others/ECJ_rulings%20Art_%206%20-%20Final%20Sept%202014-2.pdf

9 The political reasons which lay behind the drafting style employed in the Habitats Directive are explored by Andrew L.R. Jackson, *Conserving Europe's Wildlife: Law and Policy of the Natura 2000 Network of Protected Areas* (Routledge, 2018).

10 *R (Champion) v North Norfolk District Council* [2015] 1 WLR 3710.

11 [2019] EWHC 1211 (Admin).

12 Case C-127/02 *Waddenzee*, EU:C:2004:482.

13 Case C-418/04 *Commission v. Ireland*, EU:C:2007:780, para. 231.

14 See further, Yvonne Scannell, *Interaction between the EIA Directive and Articles 6.3 and 6.4 of the Habitats Directive* http://www.era-comm.eu/Cooperation_national_judges_environmental_law/module_2/11.pdf

15 The EIA process determines the process not the outcome *Kelly v. An Bord Pleanála* [2014] IEHC 400 at para. 33; Case C-420/11 *Leth v. Austria,* EU:C:2012:701, Opinion of Advocate-General at para. 42.

16 See e.g. *An Taisce - The National Trust for Ireland v An Bord Pleanála & ors* [2022] IESC 8 at [4] *per* Hogan J: "The object of the EIA Directive is itself perfectly clear, in that it seeks to ensure that the likely environmental impacts of any major project are

themselves considered and assessed before any development permission is granted, even if, as this Court has already held, 'the outcome of that examination, analysis, evaluation and identification informs, rather than determines, the planning decisions which should or may be made': *Fitzpatrick v. An Bord Pleanála* [2019] IESC 23, [2019] 3 IR 617 at 642, per Finlay Geoghegan J."

17 Foreword to Gregory Jones (ed.), *The Habitats Directive: A Developer's Obstacle Course?* (Hart, 2012).

18 Provided the correct procedures are followed the court will not intervene in the merits of the decision unless it is finds it to be "*Wednesbury* unreasonable" which it rarely, if ever does. This currently remains the position in England and Wales even in respect of the approach to the Habitats Directive see e.g. *Smyth v The Secretary of State for Communities and Local Government, Ms Elizabeth Archer Arthur and Others* [2015] EWCA Civ. 174. Whether this is correct is questioned later in this chapter. A full exposition of the role of the *Wednesbury* test in modern judicial review is outside the scope of this chapter however.

19 Article 191(2) of the Treaty on the Functioning of the European Union

> 2. Union policy on the environment shall aim at a high level of protection taking into account the diversity of situations in the various regions of the Union. It shall be based on the precautionary principle and on the principles that preventive action should be taken, that environmental damage should as a priority be rectified at source and that the polluter should pay.

20 Case C-127/02 *Waddenzee*, EU:C:2004:482.

21 At [43]-[44].

22 [2013] 3 C.M.L.R. 16; [2015] Env LR 18.

23 This genesis of this chapter is my paper of the same title delivered at University College Dublin on 1 June 2018 in the conference "Key Challenges in European Nature Law and Policy". I am most grateful to my then pupil at Francis Taylor Building, Michael Feeney for his assistance in updating the draft chapter prior to submission.

24 Case C-323/17 *People Over Wind and Sweetman v Coillte Teoranta*, EU:C:2018:244.

25 The Conservation of Habitats and Species Regulations 2017 ("the Conservation Regulations") (and the equivalent offshore SI) is "retained EU law" under the EU Withdrawal Act 2018 ("EUWA 2018") and, more specifically, "EU-derived domestic legislation". The Conservation Regulations continue to have effect in the post-Brexit world – which means in effect that the EU Habitats and Wild Birds Directives which they implement continue to have effect in England and Wales. The EU Directives are directly incorporated into the domestic framework. Regulation 9 of the Conservation Regulations provides that: "The appropriate authority, the nature conservation bodies and, in relation to the marine area, a competent authority must exercise their functions … so as to secure compliance with the requirements of the Directives." In addition, by EUWA 2018 s.5(2) the principle of supremacy of EU law continues to apply. That means that to the extent that the Conservation Regulations are inconsistent with pre-2020 domestic law, the Regulations would not prevail (although the UK Government is free to make laws overriding EU habitats law post-2020. As Ned Westaway states in *Retained EU law: Habitats and Species – Calm Before the Storm?* posted on 27 July 2021 on FTB's Environmental Law Blog ("ELB") at https://www.ftbchambers.co.uk/blogs/retained-eu-law-habitats-and-species-%E2%80%93-calm-storm: "[T]he Habitats and species protection in England and Wales therefore finds itself in a kind of holding position. The established protections to which domestic lawyers are now well-accustomed remain in force. The several judicial cases to consider in this area of law since January 2021 are clear that there has been no substantive change – see e.g. *R (Keir) v Natural England* [2021] EWHC 1059 where, among other things, Holgate J confirmed that the approach taken by Natural England to bat licensing in March 2021 "must accord with the

precautionary principle." Strictly speaking it is not quite right to say that the post Brexit current arrangements are created simply by "cut and paste" as I describe the process in the body of the chapter. As Westaway correctly points out in the same blog post: "The law is not completely unchanged." One of the reasons for treating domestic regulations as "retained EU law" was to enable amending regulations to be made under EUWA 2018 s.8 addressing "deficiencies" – effectively gaps or redundancies – caused by the departure of the UK from the EU. For the Conservation Regulations, this was primarily done by the Conservation of Habitats and Species (Amendment) (EU Exit) Regulations 2019, the intention of which was, according to the Explanatory Memorandum:" to ensure habitat and species protection and standards as set out under the Nature Directives are implemented in the same way or an equivalent way when the UK exits the EU. There is no change to policy." The EU Exit Regulations made three main types of change. First, is the nationalisation of terminology. The Natura 2000 network of sites, is now the "national site network". Similarly, "Community interest" is now "the national interest". However these are changes in name only: the basis for the assessment of significance (etc.) remains to all intents and purposes the same. Second, is the transfer of functions belonging to the Commission to the Secretary of State. For example, it now lies with the Secretary of State to give an opinion on whether imperative reasons of overriding public importance (IROPI) exist (other than public health or environmental reasons) where impacts cannot be ruled out on priority species or habitats. Reporting obligations under Articles 16 and 17 of the Habitats Directive are also brought "in house" – which in effect means self-reporting (with a role for the JNCC). Third, substantive site management obligations under Article 6(1) are retained through a new Regulation 16A. The Habitats Directive is transposed into Irish law by the European Communities (Birds and Natural Habitats) Regulations 2011 as amended (hereafter "the Habitats Regulations"). 'European sites' are defined in Regulation 2(1) of the Habitats Regulations and comprise Special Areas of Conservation (SACs) and Special Protection Areas (SPAs), at all stages of designation.

26 At [71].

27 "The Commission has an autonomous power to issue guidance documents (Article 292 TFEU referring to the Commission's power to issue recommendations) so the legislator may not impose obligation to issue guidance. Frequently, however, legislative measures contain such obligations. See, for example, Annex I, points (a) and (b), to Directive 2010/75/EU of the European Parliament and of the Council of 24 November 2010 on industrial emissions." See European Commission, "TOOL #39. GUIDANCE DOCUMENTS CONTAINING LEGAL INTERPRETATION OF EU LAW" at https://ec.europa.eu/info/sites/default/files/file_import/better-regulation-toolbox-39_en_0.pdf, at note 439.

28 In a 2016 study, Arrebola et al examined the influence of the Advocate General on the judgments of the Court, showing that the Court is approximately 67% more likely to deliver a particular outcome if that was the opinion of the Advocate General: see Carlos Arrebola, Ana Júlia Mauricio, and Héctor Jiménez, "An Econometric Analysis of the Influence of the Advocate General on the Court of Justice of the European Union" (2016) 5(1) Cambridge Journal of International and Comparative Law 82–112.

29 [2008] EWHC 1204; [2008] 2 P. & C.R. 16.

30 See further, Paul Stookes, "The Habitats Directive: Nature and Law" (Ch. 8), in Gregory Jones (ed.), *The Habitats Directive: A Developer's Obstacle Course* (Hart 2012) 140.

31 For further detailed examination of planning decisions in respect of the Thames Basin SPA See further, Simon Ricketts and Sarah Bischoff, "SANGs: The Thames Basin Case Study" (Ch. 7), in Gregory Jones (ed.), *The Habitats Directive: A Developer's Obstacle Course* (Hart, 2012).

32 The SANG mechanism is cited by Andrew Waite (Ch. 13) in Gregory Jones (ed.), *The Habitats Directive: A Developer's Obstacle Course* (Hart, 2012), as an example of the application of a concept which he calls: "The Principle of Equilibrium in Environmental Law."
33 [2003] Env LR 30; [2003] EWCA Civ 400.
34 At paras [36]-[37].
35 [2007] EWCA Civ 298 at [37].
36 It would not now have been open for Sullivan J to reason in this way given para 45 of Case C-243/15 *Lesoochranárske zoskupenie VLK 'LZ II'*, EU:C:2016:838; "[Article 6(3) of the Habitats Directive] must be read in conjunction with Article 6(1)(b) of the Aarhus Convention, an instrument which forms an integral part of the EU legal order."
37 Case C-127/02 *Waddenzee Landelijke Vereniging tot Behoud van de Waddenzee and Another v Staatssecretaris Van Landbouw, Natuurbeheer en Visserij*, EU:C:2004:482.
38 Notwithstanding AG Sharpston's own nationality she is obviously one who is excluded from this generalisation.
39 *R (Hart DC) v SSCLG* [2008] EWHC 1204.
40 AG Sharpston in C-258/11 *Sweetman*, EU:C:2012:743, said: "49. The threshold at the first stage of article 6(3) is thus a very low one. It operates merely as a trigger, in order to determine whether an appropriate assessment must be undertaken of the implications of the plan or project for the conservation objectives of the site. The purpose of that assessment is that the plan or project in question should be considered thoroughly, on the basis of what the court has termed 'the best scientific knowledge in the field'".
41 In the UK, *Wednesbury* refers to a challenge to the legality of a decision on the basis that the decision taken was irrational. The reference is to the case of *Associated Provincial Picture Houses Ltd v Wednesbury Corporation* [1948] 1 KB 223.
42 See Andrew L.R. Jackson, *Conserving Europe's Wildlife* (Routledge, 2018) 171. The book contains a detailed analysis of the negotiations of the Habitats Directive.
43 See further, Paul Stookes, supra n 18, at p. 148.
44 [2015] EWCA Civ. 174; [2015] P.T.S.R. 1417 at [68]–[75]
45 At [74].
46 [2015] Env. L.R. 28 at [72]–[74].
47 [2015] 1 W.L.R. 371. See e.g. Lord Carnwath JSC at [42].
48 [2017] UKSC 37.
49 At [23].
50 [2018] EWCA at [41].
51 *St Modwen Developments v. SSCLG* [2017] EWCA Civ. 1643 at [7].
52 [2013] NIQB 30.
53 At [91].
54 See also *Smyth v SSCLG* [2015] EWCA Civ 174; [2015] P.T.S.R. 1417.
55 In *Meadows v. Minister for Justice* [2010] IESC 3, Denham J in the Irish Supreme Court recognised that the test for irrationality still applies to the decisions of An Bord Pleanála:

> The decision in *O'Keeffe v. An Bord Pleanála* related to a specialised area of decision making where the decision maker has special technical or professional skill. A court should be slow to intervene in a decision made with special competence in an area of special knowledge. The *O'Keeffe v. An Bord Pleanála* decision is relevant to areas of special skill and knowledge, such as planning and development.

56 Michael Fordham, *Judicial Review Handbook* (7th ed., Hart 2020) 49.1.
57 *R (FZ) v Croydon LBC* [2011] EWCA Civ. 59; [2011] PTSR 748 at [4].
58 [2003] EWCA Civ 1408; [2004] Env LR 391 at [17].
59 [2014] IEHC 557.

60 Eirgrid had applied to the Board for a declaration that the construction of new higher pylons to replace existing pylons along the same route corridor of a 110 kV electricity line, the replacement of the line itself, and the installation of a temporary line while the works were being carried on, are exempted development. Eirgrid submitted a screening report which identified a potential indirect impact on water quality within the SAC, but concluded that such impact was unlikely to occur due to the localised and restricted nature of the works and on the basis that "best practice construction methods" would be employed. The applicants for judicial review had however submitted a report that argued that indirect impacts could not be ruled out. They challenged An Bord Pleanála's decision amongst other things on grounds that it had acted unlawfully in taking into account the proposed mitigation in the screening assessment. An Bord Pleanála had screened out the need for an AA, concluding that there was no reason to suggest that the works proposed would have any significant effect on the nearby Rye Water Valley SAC, in view of its qualifying interests. In this context, the Board overruled its inspector, who had recommended that an AA should be carried out, due to residual doubts and a perceived lack of detail in the evidence before the Board. It is to be noted that in *Hart* the Secretary of State had also overruled the view of his inspector who had concluded that an AA should have been required, as to which see further Peter Scott, "Appropriate Assessment: A Paper Tiger?" (Ch. 6) in Gregory Jones (ed.), *The Habitats Directive: A Developer's Obstacle Course* (Hart, 2012), p. 115.

61 [2014] IEHC 400.

62 Alice Whittaker, "Pylons, Habitats and Best Practice Construction Methods," "Planning & Environment Case Update," September 2014, Philip Lee Solicitors. http://www.philiplee.ie/wp-content/uploads/2015/03/Environmental-and-Planning-Case-Update-September-2014.pdf

63 [2015] EWCA Civ 174; [2015] P.T.S.R. 1417.

64 The preliminary reference procedure is set out in Article 267 of the Treaty on the Functioning of the European Union.

65 [2017] IEHC 171.

66 The guidance given by the CJEU does not require them to give a clear indication of what their own opinion on the matter might be. See e.g. Recommendations 2019/C 380/01 dated 8 November 2019. Useful guidance is also provided in the Fair Trials – Preliminary Reference Toolkit, March 2019. https://www.fairtrials.org/sites/default/files/CJEU%20preliminary%20reference%20Toolkit_0.pdf

67 [2007] EWCA Civ 298.

68 At [35].

69 At [36].

70 At [38].

71 At [37].

72 At [39].

73 CJEU, Consolidated version of Protocol (No 3) on the Statute of the Court of Justice of the European Union, Article 20, para. 5.

74 The Court of Appeal composition was particularly distinguished with great public law expertise. Both Sales LJ and Richards LJ had been First Junior Treasury Counsel, known colloquially as the "Treasury Devil" representing Her Majesty's Government in the civil courts. Sales LJ and Kitchen LJ were both subsequently promoted to the Supreme Court.

75 Case C-387/15 *Orleans and others*, EU:C:2016:583.

76 C-164/17 *Grace and Sweetman v An Bord Pleanála*, EU:C:2018:593.

77 The CJEU concluded that

> [57]. … where it is intended to carry out a project on a site designated for the protection and conservation of certain species, of which the area suitable for providing for the needs of a protected species fluctuates over time, and the temporary

or permanent effect of that project will be that some parts of the site will no longer be able to provide a suitable habitat for the species in question, the fact that the project includes measures to ensure that, after an appropriate assessment of the implications of the project has been carried out and throughout the lifetime of the project, the part of the site that is in fact likely to provide a suitable habitat will not be reduced and indeed may be enhanced may not be taken into account for the purpose of the assessment that must be carried out in accordance with Article 6(3) of the directive to ensure that the project in question will not adversely affect the integrity of the site concerned; that fact falls to be considered, if need be, under Article 6(4) of the directive.

78 Joined Cases C-293/17, C-294/17, EU:C:2018:882.
79 Case C-461/17 *Holohan v. An Bord Pleanála*, EU:C:2018:649.
80 [2019] IEHC 84.
81 [2019] IEHC 84, [68].
82 [2019] IEHC 84, [104].
83 [2019] IEHC 84, [148].
84 *Heather Hill Management Company v An Bord Pleanála* [2019] IEHC 450 and *Uí Mhuirnín v Minister for Housing Planning and Local Government* [2019] IEHC 824.
85 [2019] IEHC 824.
86 [2019] IEHC 824, [15].
87 Elizabeth Dunn, 'Habitats Directive: What Mitigation Can Be Taken into Account?' *Burges Salmon* (4 September 2018), https://www.burges-salmon.com/news-and-insight/legal-updates/habitats-directive-what-mitigation-can-be-taken-into-account/
88 Gladman Land, ''People Over Wind' Ruling Implications on HRA', June 2018, Issue 37, http://www.gladmanland.co.uk/article/people-over-wind-ruling-implications-on-hra/
89 Tom Edwards, 'CJEU: Mitigation Should Not Be Considered at Habitats Screening Stage' 8 May 2018, Out-Law Analysis, https://www.pinsentmasons.com/out-law/analysis/cjeu-mitigation-habitats-screening-stage
90 Paragraph 175 of the then 2012 NPPF stated that, "the presumption in favour of sustainable development does not apply where development requiring appropriate assessment under the Birds or Habitats Directives is being considered, planned or determined." This paragraph became paragraph 177 in the 2018 version subject only to minor cosmetic changes. In February 2019 paragraph 177 of the NPPF was amended to state that the presumption in favour of sustainable development would not apply where there are effects on a habitats site unless an appropriate assessment concludes that the plan or project will not adversely affect the integrity of the habitats site. Previously, the presumption would not apply merely where an AA had been triggered. Although this change was generally said to have been prompted by the *People Over Wind* ruling the original text was open to the criticism that there was no rational basis for disapplying the presumption in favour of sustainable development merely because an AA had been triggered if the AA ultimately concluded that negative impacts would not occur.
91 [2019] EWHC 1211 (Admin).
92 [2012] UKSC 44; [2013] PTSR 5.
93 [2015] 1 WLR 3710.
94 [2019] EWHC 2001 (Admin).
95 The Planning Inspectorate, 'Consideration of avoidance and reduction measures in Habitats Regulations Assessment: People Over Wind, Peter Sweetman v Coillte Teoranta', PINS Note 05/2018, 9 May 2018, to guide local plans inspectors.
96 [2019] Env. L.R. 9.
97 [2019] EWCA Civ 1562.
98 [2021] EWCA Civ 592.

99 [2019] EWHC 3505 (Admin) at [91].
100 [2019] EWHC 3505 (Admin), at [91].
101 [2021] EWCA Civ 592, [95].
102 [2019] EWHC 1974 (Admin) at [78]–[80].
103 'Consequences of the United Kingdom's withdrawal from the European Union for the Court of Justice of the European Union', Court of Justice of European Union, Press Release No. 10/20. In accordance with the terms of the withdrawal agreement, the Court of Justice continued to have jurisdiction in any proceedings brought by or against the UK before the end of the transition period, which was set as 31 December 2020. It also continued to have jurisdiction to give preliminary rulings on requests from courts and tribunals of the UK made before the end of the transition period.
104 Findings and recommendations with regard to communication ACCC/C/2013/90 concerning compliance by the United Kingdom of Great Britain and Northern Ireland. Adopted by the Compliance Committee on 26 July 2021.
105 A recent Green Paper states that "the UK Government wants to fundamentally change the way the assessments under Habitats Regulations work": Department for Environment, Food and Rural Affairs, 'Nature Recovery Green Paper: Protected Sites and Species', March 2022. As part of this process, the Habitats Regulations Assessment (HRA) Review Working Group put forward a spectrum of "solutions", amongst the less radical of which is a suggestion to allow "consideration of the mitigation hierarchy earlier in the assessment process": Habitats Regulations Assessment Review Working Group summary of findings, undated, https://consult.defra.gov.uk/nature-recovery-green-paper/nature-recovery-green-paper/supporting_documents/Background%20Doc%201%20%20HRA%20Review%20Working%20Group%20%20Summary%20of%20Findings.pdf. The Retained EU Law Bill — which requires all UK government departments to either repeal or reform all EU-derived law still in force by the end of 2023 — is yet to be passed at the time of writing and is the subject of controversy. But in any event, as the legislation currently stands in terms of the Habitats Regulations, the approach of the UK courts post Brexit can be gleaned from the Court of Appeal's judgment in *R (Wyatt) v Fareham BC & Natural England* [2022] EWCA Civ. 983. In that case the Court noted that it was bound by the Supreme Court's judgment in *R (on the application of Friends of the Earth Ltd and others) v Heathrow Airport Ltd* [2020] UKSC 52 and applied the Wednesbury test to the adequacy of the AA; holding that assumptions which were not "best scientific evidence" were sufficiently precise to satisfy the precautionary approach used for assessing the adequacy of the AA. This may be contrasted with the emerging more intensive judicial scrutiny applied to screening for AA applied by the Irish courts who see it as more of a matter of a gateway to jurisdiction, see Á Ryall, 'Reviewing Science & Law in Member States' Courts: Enforcement of the Habitats Directive in Ireland' (REALaw.blog, 16 December 2022) https://realaw.blog/2022/12/16/reviewing-science-law-in-member-states-courts-enforcement-of-the-habitats-directive-in-ireland-by-a-ryall/

7

SUPPORTING NATURE CONSERVATION THROUGH AN ECOSYSTEM SERVICES POLICY

Ian Hodge

Introduction

Nature conservation is facing particular challenges. Current policy approaches have not achieved the environmental standards to which commitments have been made and the processes of climate change are further compromising our capacity to attain those standards. Approaches to and priorities for nature conservation practice have changed since the core elements of nature conservation policy were introduced in the UK in the 1940s and in European legislation under the Birds and Habitats Directives since the late 1970s. At the same time, habitats and species have often been harmed by the policies implemented under the Common Agricultural Policy (CAP) particularly under Pillar 1, although against this, nature conservation activities have been supported under Pillar 2 of the CAP. This has provided a major source of funding for nature conservation, delivered through agri-environment schemes implemented under the Rural Development Regulation. Payments are available on a voluntary basis to farmers who adopt certain agricultural practices to compensate for income foregone and additional costs of entering into fixed-term environmental contracts.

Brexit provides the UK with a once in a generation opportunity to develop a comprehensive and more systematic approach to rural land policy. Since the referendum, a variety of initiatives have been implemented, in England, under the Agriculture Act 2020 and Environment Act 2021. Most significantly, the direct payments paid under the CAP are being phased out over a seven-year transition period. In their place, agricultural policy will be centred around an Environmental Land Management scheme. This includes three main elements: the Sustainable Farming Incentive (SFI) available to all farmers, paying for actions to manage land in an environmentally sustainable way; Local Nature Recovery (LNR), paying for actions that support local environmental priorities for nature

DOI: 10.4324/9780429299100-7

recovery; and Landscape Recovery (LR) supporting the delivery of landscape and ecosystem recovery through long-term, large-scale land use change projects. These initiatives offer valuable elements for future policy but fail to offer a coherent whole.

In this chapter, we first briefly outline the way in which a British Ecosystem Services Policy (BESP) might operate at a national scale to replace the role of the CAP. This would extend land policy to take better account of the full range of ecosystem services. We then consider the changing context within which nature conservation is set and the implications for nature conservation policies. Nature conservation policy needs to respond particularly to the challenges of climate change, to do more to support nature conservation in the wider countryside and to integrate nature conservation activities better with other rural land values and practices. This indicates a need to move further beyond the site-based policy that has dominated in the past but we lack the systematic governance arrangements through which to implement a new approach. We assess the way in which such an ecosystem services policy can contribute towards the development of new nature conservation policies.

A British ecosystem services policy

The core aim of an ecosystems services policy is to maximise the long-term social value derived from rural land through the maintenance of natural capital and the delivery of a broad range of ecosystem services. A broad outline of how a British Ecosystem Services Policy (BESP) might operate has been proposed by Gawith and Hodge.[1,2] The policy would support delivery of ecosystem services (ESs) by both landholders and other organisations and individuals who can contribute effectively to the delivery process. Provision of ESs would be implemented by developing Payment for Ecosystem Services (PES) schemes where feasible but primarily it would be implemented through a series of public procurement funds administered at national and local levels. National funds would address international commitments as well as nationally determined targets, while devolved local funds would address demands from a local perspective. Funding to support this would be provided from the direct payments previously paid under the Common Agricultural Policy, which, as noted, are being phased out over a period of time.[3] Landholders and others, individually or, preferably, collectively and in partnership with other stakeholders, could tender to deliver services, potentially providing a portfolio of marketed and non-marketed ecosystem services. Organisations operating funds would monitor the outcomes, costs revealed through tendering and the effectiveness of alternative delivery methods, implementing an adaptive management approach.

A potential architecture of the main elements of a BESP is illustrated in Figure 7.1. National procurement funds would be funded through national taxation. They would procure ES from groups of farmers or NGOs who would in turn procure ESs from individual land holders, either as members of these

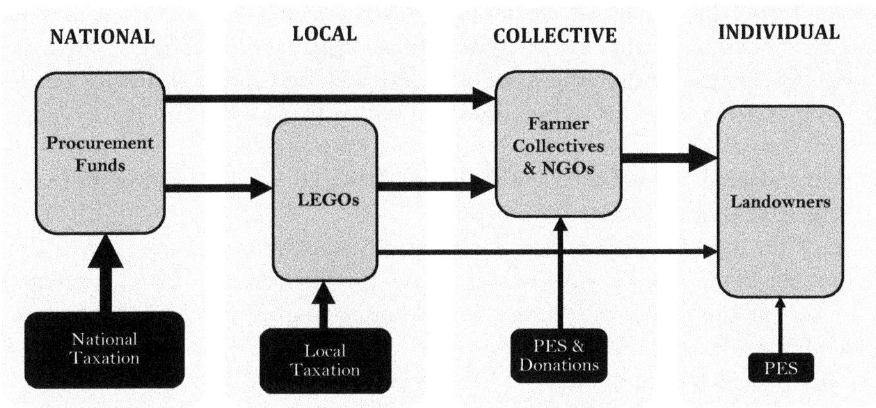

FIGURE 7.1 Potential architecture for a British Ecosystem Services approach.
Source: Adapted from Fig. 2 of Gawith and Hodge, 2019, n 2.

groups or through environmental contracts. There would also be a devolved tier of procurement from Local Environmental Governance Organisations (LEGOs) which could support delivery of ES both through collective organisations and directly from individual landholders. LEGOs could be funded through a combination of national and local taxation. Information on the outcomes of these activities would be fed back to procurement funds and inform future funding allocations.

Shifting sands of nature conservation

Nature conservation policies are not working well. A broad range of publications have pointed to continuing losses of biodiversity. The State of Nature Report[4] shows that in the UK, notwithstanding some gains, the abundance indicator for 697 terrestrial and freshwater species shows a statistically significant decline in average abundance of 13% between 1970 and 2016. Since 1970, the indicator of abundance for 214 priority species has declined by 60%, and between 2011 and 2016 by 22%. The UK index of 19 farmland bird species declined by 56% between 1970 and 2017.[5] Recent attention has been drawn to the widespread loss of pollinating insects in Britain[6] and only 39% of Sites of Special Scientific Interest in England are considered to be in 'good status'.[7] The UK review of progress towards the Aichi Targets[8] concluded that progress was insufficient to achieve Targets 1, 3, 4, 5, 6, 7, 8, 9, 10, 12, 13, 14 and 15 by the deadline set.

The causes for these losses are complex. A study of the drivers of change across 400 species sampled from a broad range of taxonomic groups in the UK[9] concluded that species change between 1970 and 2012 has been most strongly impacted by intensive management of agricultural land and by climatic change. However, the solution is not simply to stop agricultural land uses. In a cultural landscape, such as in the UK, and characteristic of many areas within Europe, conservation values in farmland and woodland generally depend on the continuation of particular agricultural and forestry management practices.[10] These are often more extensive than the most profitable activities and may involve mixed farming systems or more labour-intensive methods. Thus nature conservation policy requires the establishment of incentives for particular forms of land management adapted to conditions within particular localities, both within protected areas as well as in the wider countryside.

In 2018, the UK Government[11] set out a bold ambition in its 25 Year Environment Plan. The Plan pledged to 'leave the environment in a better state than we inherited it by protecting wildlife, improving air and water quality and curbing the scourge of plastic in the world's oceans'. The plan included a range of objectives including those for 'Thriving plants and wildlife':

> We will achieve a growing and resilient network of land, water and sea that is richer in plants and wildlife. On land and in freshwaters, we will do this by:
>
> - restoring 75% of our one million hectares of terrestrial and freshwater protected sites to favourable condition, securing their wildlife value for the long term;
> - creating or restoring 500,000 hectares of wildlife-rich habitat outside the protected site network;
> - taking action to recover threatened, iconic or economically important species of animals, plants and fungi;
> - increasing woodland in England in line with our aspiration of 12% cover by 2060: this would involve planting 180,000 hectares by end of 2042.

Further, the Environment Act 2021 introduces a target to halt the decline in species abundance by 2030 and mandates that, in addition, at least one other long-term biodiversity target must be set. Consultation on these targets was undertaken in 2022. While strong on ambition, it is less clear how these targets are to be delivered in practice. This raises a wide range of questions. For instance:

- How would the location of interventions, such as to create habitat, be determined within particular local contexts?
- How would the investments in conservation be secured in the longer term?
- How would interventions to promote 'thriving plants and wildlife' be co-ordinated with other interventions to sequester carbon or to improve water quality?
- How would interventions be assessed and selected so as to promote value for money?
- How would the values of different stakeholders be taken into account in determining the approaches to be taken?
- How would the progress be monitored and adjusted in response to changes in the context and objectives over time?

Clearly, policy design, especially in implementing a 25-Year Plan and meeting other long-term targets, and recognising the significance of path dependency in making policy changes difficult, needs to take into account the likely future conditions within which policy will be applied. In order to realise such long-term ambition, policies need to take account of the changes in thinking about nature conservation, moving beyond the pre-existing focus on site designation and protection. More recent discussions of nature conservation have emphasised larger-scale interventions. This is particularly represented by the arguments of the Lawton Committee[12] for conservation to be 'more, bigger, better and joined up'. And a key factor for conservation will be climate change. This has the potential to challenge some of the basic assumptions of nature conservation.[13] One particular issue will be an increased acceptance of habitat change in contrast to the previous emphasis on protecting particular species within particular locations. Strategies are likely to be more adaptive, such as through dynamic protected areas, assisted migration, and the expansion of linked protected networks. Isaac et al.[14] focus on the delivery of resilient ecological networks, although they accept that there is a lack of clarity about what such a network would look like. They propose five potential targets for delivering 'bigger, better, more and joined wildlife sites': improving the condition of protected areas, improving the condition of landscapes that are not currently protected for nature conservation but have broader roles, such as national parks, increasing the area of habitats under long-term protection for nature conservation, establishing large habitat areas by creation and/or restoration, and improving the quality and extent of habitat connectivity.

Governance requirements for future nature conservation actions

In light of Brexit, new governance arrangements are needed in order to meet nature conservation commitments and deliver the objectives of the 25 Year Environment Plan. This requires governance mechanisms:

- to determine appropriate management interventions that both meet national objectives and reflect local priorities;
- to promote co-ordinated land management at a scale in excess of individual landholdings or ownership units;
- to integrate the delivery of nature conservation with the delivery of other ecosystem services;
- to develop long-lasting arrangements that secure the environmental gains indefinitely into the future;
- to adapt to new information and changing circumstances over the 25-year period.

We address each of these issues in turn.

Responding to national and local priorities

Drivers of nature conservation activities may arise at national or local levels. National governments commit to international agreements as well as setting national priorities. National parks, for example, represent a national commitment to the environmental quality of a particular locality. On the other hand, some concerns reflect more local preferences, perhaps to maintain a local landscape or provide public access. These local preferences are unlikely to be identified by national government. This implies a need for polycentric governance with capability for decision-making at multiple levels. In the proposal for a BESP, this is simply represented by a national and a local level.

Governance arrangements for environmental decision-making at a national level are well established but this is not generally the case at more local levels. This implies a need to develop Local Environmental Governance Organisations (LEGOs) that can represent local priorities in the delivery of nature conservation and ecosystem services more generally. The introduction of Local Nature Recovery Strategies under the Environment Act 2021 will do something to identify and represent priorities, and collaboration amongst local stakeholders is being developed by Local Nature Partnerships, such as in Devon or Cambridgeshire. But these initiatives do not establish powers by which the strategies are to be implemented. We might envisage a LEGO as a custodian of local environmental values to secure and promote long-term local environmental sustainability. They could be run by a democratically appointed board of trustees tasked to secure the best long-term social value of ecosystem services to the local community.

National priorities might be incentivised through national government regulations or payments directly to landholders or they could be implemented indirectly through incentives for LEGOs to pursue national objectives. LEGOs would be funded from a combination of national and local sources, the balance between these representing both the national significance of the local area as well as equity concerns.

The scale of area across which LEGOs would operate would aim to represent areas of common interests where ecosystem management can be enhanced through spatial co-ordination, such as catchments, in relatively homogeneous environments or areas that share a common culture or outlook. At the same time, areas need to be of sufficient scale to justify professional administration and democratic input. A British National Park suggests some elements of the sort of governance arrangement that might be followed.[15] In practice, scale of area is likely to vary from region to region. Similarly, the intensity of LEGO activities would also vary spatially, depending on the social value of the environmental assets present and the level of conflict over them.

Delivering landscape-scale conservation

Historically, conservation practice has tended to be implemented at the scale of the individual ownership unit or protected area. Agri-environment contracts are predominantly agreed between a national government agency and an individual farmer. But, as noted above, it is now widely accepted that conservation can be more effective if land management interventions can be co-ordinated across larger spatial areas. This requires mechanisms to co-ordinate decisions taken by different land management agents within those areas.

This can be achieved by voluntary co-ordination, collective agreements or by control by a single conservation entity. A range of initiatives have been taken, led by non-governmental organisations, to promote nature conservation across larger areas of land.[16] These include the Living Landscapes initiative by the Wildlife Trusts and Futurescapes initiative by the Royal Society for the Protection of Birds. Large-scale conservation initiatives are primarily led by private conservation NGOs and implemented through partnerships in a variety of types of arrangement with other private and public organisations. They now cover a substantial proportion of the country. The policy relies heavily on voluntary initiatives supported by small amounts of money and such initiatives have tended historically to draw to a large extent on agri-environment funding provided under the Common Agricultural Policy (CAP). Alongside the NGO initiatives, more recently there has been an emergence of a substantial number of farmer clusters in England where groups of farmers develop and implement an overall plan for their conservation activities.[17]

Voluntary co-ordination can be successful. However, more radical changes to land uses and management and more secure conservation outcomes are likely to require a more formal approach. This can be secured through farmer associations or cooperatives or land trusts. Farmer cooperatives have been widely adopted in implementing agri-environment schemes in the Netherlands.[18] Collective contracts with groups of landholders have the advantage of allowing detailed management decisions on an overall plan. This includes agreement on the contributions being made and the payments being made by the individual members of the group. They will each have good information about their respective roles and what constitutes fair compensation for actions taken. They will also have a collective interest in monitoring and enforcement. Alternatively there could be broader partnership arrangements. Wider partnerships can include non-farming collaborators who may provide ecological or research expertise, capacity to monitor progress or voluntary labour. This can bring a wider range of resources and expertise into the project but at the same time complicates the internal arrangements and raises transaction costs. With any of these approaches, there is a need for trust amongst the participants and it will take time to develop the necessary social capital alongside the more formal contractual relationships. Control by a single agent can more easily implement radical, coherent and long-term changes to a land management system. Most obviously this applies where a sufficiently large area of land is under single ownership. A good example of this is the rewilding initiative on the Knepp Estate in West Sussex[19] or on land owned by the National Trust at Wicken Fen in Cambridgeshire.

Some support for this type of approach can be provided through the Landscape Recovery element of the Environmental Land Management scheme (which, as noted above, replaces the CAP). Landscape Recovery is aimed at large-scale projects (500–5,000 ha) over the longer term (20 years+). Each project will have bespoke arrangements, individually negotiated with central government by a single legal entity on behalf of the other participants. Projects will be funded by a mix of public and private finance. Applications for pilot projects have been invited in 2022, focussing on recovering threatened native species and restoring streams and rivers. But projects will be encouraged to deliver other ecosystem services in addition. However, these projects lack local governance and it is not clear to what extent they will respond to local priorities, whether as represented in Local Nature Recovery Strategies or in other local policies.

Integrating provision across ecosystem services

Nature conservation policies tend to be legislated for and implemented independently of other policies by specialist nature conservation agencies. Protected areas are designated on the basis of their nature conservation status and managed to protect and promote them. However, there is evidence to suggest complementarities and trade-offs in the provision of multiple ecosystem services from particular areas of land.[20] This might be seen in terms of economies of scope, where gains can be achieved by producing two or more separate outputs together, in contrast to economies of scale where unit costs of production fall as the scale of production increases. Thus, sites that are of value for nature conservation may also have the capacity to deliver a range of other ecosystem services, such as landscape, carbon storage or public access. A system of site selection that focusses only on interventions directed at nature conservation characteristics may thus miss out on the potential to deliver co-benefits.

It is likely that the potential to deliver combinations of multiple ecosystem services at a particular location will be best known to the landholder at that location rather than to any single government agency. Conservation scientists may well have good information on the capacity for particular sites to provide particular nature conservation benefits. This tends to focus on individual species, and the assessment of a site becomes more complex in the context of considering multiple species. But the position with regard to the delivery of multiple ecosystem services is more complex again. This requires judgement not simply of the potential to deliver benefits but, significantly, information is also required on the opportunity costs, what is foregone in terms of lost production opportunities, and, perhaps hardest to know of all, the knowledge, skills and preferences of the individual land manager herself.

An objective of an ecosystems services policy is thus to incentivise landholders to seek out opportunities to manage their properties for the delivery of a portfolio of multiple benefits, including marketed agricultural products. Landholders may, of course, benefit from guidance from specialist ecologists as to the likely conservation potential of their property. Incentives might be provided through separate funds procuring individual ecosystem services, such as carbon sequestration or flood protection, or by funds procuring bundles of different ecosystem services from single landholders. The objective for cost-effectiveness, in either case, is for the fund only to pay for the marginal increment in ecosystem service output stimulated by the operation of the payment. Defra[21] has discussed the options of layering, single ecosystem services procured by a separate fund, and bundling, sets of ecosystem services procured by a single fund. More work is required to explore these alternative approaches, which have implications for the appropriate balance between national and more local approaches towards procurement.

Securing environmental gains in the long term

While protected area designation operates over an indefinite time period, funding to support nature conservation actions, such as under an agri-environment scheme or a project funded by the National Lottery Heritage Fund operates on a voluntary basis over a fixed and limited period of time, often for about five years. The consequence is that at the end of this period, or perhaps at breakpoints within an agreement, either party to the agreement can have the opportunity to withdraw. This presents a risk that any conservation benefits built up over the contract period can be lost if the contract is not renewed and the land is put back into more intensive agricultural use. The risk of this outcome depends on the circumstances at the time when the contract ends. For instance, if commodity prices are relatively high and the amount paid in agri-environment schemes is no longer seen to cover the costs of complying with the conditions, then participants might be expected to not renew.

There are various options that can at least in part mitigate this risk. Land might be purchased by government or by a conservation NGO. A conservation trust can have the delivery of the nature conservation value as a permanent objective and can cross-subsidise land management from other sources of income, such as membership fees or donations. Alternatively, a conservation covenant might be used to establish a legal control over future land use options without the acquisition of the freehold.[22] Provision for conservation covenants is made in the Environment Act 2021 and they are expected to come into force in September 2022. These options are likely to be relatively expensive. This is clearly the case with land purchase, but farmers are typically reluctant to accept long-term or permanent constraints over the way in which their property can be used. They would look for a substantially larger payment in order to be willing to accept a permanent covenant. The exception could be for landowners who are seeking to secure conservation values on their land, perhaps after they cease to manage the property themselves. In this case, they may be willing to donate a voluntary conservation covenant without payment in order to secure future environmental quality.

Adapting to changing circumstances

In practice, the outcomes to intervention into ecosystem processes are uncertain. In this context, governance should be adaptive and applied at a landscape or catchment scale. The interrelationships and feedbacks amongst ecosystem functions and values of the services are imperfectly understood. Ecosystem management involves trade-offs with different actions benefiting different taxa and ecosystem functions, and actions can take decades to become effective.[23] Outcomes are vulnerable to unpredictable changes in external factors and so the consequences of ecosystem interventions cannot be predicted with certainty. It is thus argued that sustainable management should be adaptive and focus on building the resilience of the system.[24] Adaptive management[25] recognises this context and argues that management cannot set clear objectives but rather operates on an iterative basis, seeing interventions as experiments to generate information to feed back into future decisions. The aim of ecosystem management may thus be to build the resilience of the system against unknown future shocks, such as through the maintenance of functional redundancy to underpin service provision, rather than to seek to achieve a predetermined output.

Further, the system being managed is not simply the 'natural' environment but rather is a complex mix of natural environment and human factors combined together in a social-ecological system (SES).[26] Taken together this indicates a role for the adaptive governance of social-ecological complexity[27] defined as 'a process by which institutional arrangements and ecological knowledge are tested and revised in a dynamic, ongoing, self-organized process of learning-by-doing'.[28] The approach has close parallels with that based on landscape services.[29] The critical question concerns the appropriate governance arrangements that can deliver the required management processes. Lubell[30] has argued that determining 'which institutional structures work best in different situations is one of the most important unresolved questions in the policy sciences'. This is a fundamental issue to be addressed in the development of a British Ecosystem Services Policy.

There is perhaps a more fundamental contradiction between the short-term project funding generally available to fund nature conservation and the long-term processes of ecological restoration or rewilding. Restoration projects are likely to require relatively long periods of time before they show results and those running the projects may not have a clear idea of what changes in biodiversity will occur when the project is started. This presents a challenge for the governance of large-scale conservation.[31]

Organisations funding projects seek to ensure accountability and good financial management by requiring applicants for funds to specify the outputs that will be delivered over a defined, and generally relatively, short period of time, often with milestones to be delivered along the way. In contrast, given the complexity of and uncertainty around ecosystem processes, ecological restoration will seek to adopt a long-term adaptive approach without a clear target end state and adjusting management over time as more information about the process is revealed. On the other hand, ecologists argue that this means that it is not possible to know how to deliver predetermined outcomes and that governance should be adaptive, long-term and implemented through networks of stakeholders. Both of these approaches have strengths. Project management provides control and accountability to funders. Adaptive governance recognises complexity and provides for long-term learning, building networks and adaptive responses. Some compromise may be possible through the development of longer-term programme funding where funding organisations build relationships with those running projects, where the expectation is that funding will continue subject to satisfactory performance, and where performance is assessed not simply in terms of the delivery of outputs but more of the quality and coherence of project management. This raises challenging questions about project appraisal methodologies and monitoring that require further research.

There is also a potential conflict here between the approach that is taken to the designation of protected areas and the adoption of a more adaptive approach. Sites are designated against a well-defined set of criteria, specifically for Sites of Special Scientific Interest, these are: Typicalness, Fragility, Size, Diversity, Naturalness, Rarity, Ecological coherence, Potential value, and Recorded History. Individual sites are thus selected against criteria based on species composition that identify unique characteristics and management plans are drawn up against those criteria. It then becomes very difficult to alter the objectives or the management approach over time to deliver different types of habitats or different species. If the original objectives of designation are no longer the highest priority for a particular site, then it becomes difficult to justify designation at all. There is therefore considerable inertia deterring change away from a site's original objectives and management plan when in a more dynamic and adaptive approach the priorities and the management of the site might be expected to change. Where sites are on privately owned land, owners might be expected to argue that if the original significance of a site no longer applies, then it should no longer be designated. This gets to the fundamental principles of protected area governance and requires a more critical debate than has been achieved so far.

A balance between alternative paradigms for nature conservation

Integrating nature conservation into a broader ecosystem services policy approach shifts the paradigm of nature conservation from an approach based on selecting and defending the sites that provide the best examples of species and habitats further towards a more flexible, adaptive approach; more towards a resilient network. Some of the key differences between these approaches are summarised in Table 7.1.

TABLE 7.1 Alternative paradigms for nature conservation

	Site-based conservation	*Ecosystem services policy*
Approach	Site selection and defence	Adaptation and resilience
General objective	Protecting 'best' species and habitats	Resilient network
Intended outcomes	Protection of targeted species and habitats	Highest 'social value' ecosystem
Spatial target	Separate sites	Environmental systems
Methods	Designation and active intervention	Managing 'natural' systems
Actors	Specialist conservation managers	Multidisciplinary partnerships
Monitoring	Species counts	Measures of resilience
Funding	Separate conservation budget	Blended ecosystem services budget
Risks	Targeted species cannot survive	New habitats are of less 'value' than pre-existing

Clearly, both are simplifications of alternative approaches to conservation policy and can be seen to be implemented alongside each other in practice. The argument here is not that we should simply move from one to the other. Identification of key sites and action to support their conservation value will continue to be important. Rather it is that there can be potential gains from some degree of shift further beyond a site-based focus towards a wider ecosystem-based focus. The potential gains are that a larger proportion of the countryside can be managed taking nature conservation objectives into account, that the net cost of nature conservation actions can be reduced to the extent that there are synergistic relationships or economies of scope in the delivery of different ecosystem services, and that species populations can be more resilient against external shocks such as extreme weather events or invasion by alien species. Against this is the risk that the best examples of particular habitats may get less protection and less active intervention, with the prospect that species may be lost from those sites. Of course, it would still be possible to adopt the same intervention approaches towards selected protected areas as are applied now, and it may be the case that these species would not survive within these locations whatever approach were to be adopted.

Conclusions

The UK Government has demonstrated a clear ambition for nature conservation, both internationally by signing up to the Aichi targets and domestically through the 25 Year Environment Plan and Environment Act. But these targets are not being delivered through current policies, more needs to be done. Two principal impediments to progress are represented by the challenges posed by climate change and the political realities of limits on public funding, government's willingness to drive change and institutional capacity within government. Brexit has provided a unique opportunity to redesign rural land policy in order to address these issues. We have argued elsewhere for the introduction of a British Ecosystem Services Policy that can offer a clearer and more coherent approach to maximising the long-term social value of rural land. The argument in this chapter is that this approach can accommodate the emerging shift in nature conservation policy towards the delivery of a resilient ecological network. This would integrate nature conservation goals into a wider framework of ecosystem service delivery.

There is clearly much that needs to be done in order to realise this broader vision. Recent legislation has introduced elements into policy that can contribute towards this vision. But there is a lack of an overall coherent plan for their integration. There is a need for new governance structures that can effectively determine values of alternative conservation outcomes and establish incentives for their delivery. Collaborative working, both within decision-making processes and on the ground in larger-scale co-ordinated rural land management, will both become more important. This requires the development of new forms of social capital as well as legal arrangements that can provide trust and security for new working relationships. We need to understand better the potential complementarities and conflicts amongst the provision of ecosystem services and how this can feed into incentives for rural land managers to develop portfolios of ecosystem service provision that capitalise on the complementarities within particular local contexts, are responsive to the assembly of better information over time, and secure the increments to natural capital that are achieved in the longer term.

This indicates substantial changes from current practice. It will not be achieved immediately or without a degree of disruption. So we need a clear vision of what is being sought in the longer term and a strategy by which to achieve it. The process should be evolutionary to avoid damaging disruption to natural or social capital as the adjustments are made. The principle of a long-term ambition has been established through the 25 Year Environment Plan but we need parallel changes in governance arrangements that can bring it to fruition. As expressed by the Secretary of State for Environment,[32] we should seize the 'unfrozen moment' offered by Brexit to launch our approach to rural land policy in a new direction.

Notes

1 David Gawith and Ian Hodge, 'Envisioning a British Ecosystem Services Policy' (CSaP and Department of Land Economy 2017) http://www.csap.cam.ac.uk/links/13/4840/.

2 David Gawith and Ian Hodge, 'Focus Rural Land Policies on Ecosystem Services, Not Agriculture' (2019) 3(8) *Nature Ecology & Evolution* 1136–1139.

3 Defra, 'Health and Harmony: The Future for Food, Farming and the Environment in a Green Brexit' (2018) Cm 9577.
4 Daniel B. Hayhow et al, *State of Nature 2019* (The State of Nature Partnership 2019).
5 Defra, 'Wild Bird Populations in the UK, 1970–2017' (2018).
6 Gary D. Powney et al, 'Widespread Losses of Pollinating Insects in Britain' (2019) 10 *Nature Communications* 1018.
7 https://designatedsites.naturalengland.org.uk/ReportConditionSummary.aspx?SiteType=ALL (accessed 19 May 2022).
8 JNCC, 'Sixth National Report to the United Nations Convention on Biological Diversity: United Kingdom of Great Britain and Northern Ireland Overview of the UK Assessment of Progress for the Aichi Targets' (JNCC 2019).
9 Fiona Burns, et al, 'Agricultural Management and Climatic Change Are the Major Drivers of Biodiversity Change in the UK' (2016) 11 *PLOS ONE* e0151595.
10 Ian Hodge, Jennifer Hauck and Aletta Bonn, 'The Alignment of Agricultural and Nature Conservation Policies in the European Union' (2015) 29 *Conservation Biology* 996.
11 HM Government, 'A Green Future: Our 25 Year Plan to Improve the Environment' (2018).
12 John H. Lawton et al, *Making Space for Nature : A Review of England's Wildlife Sites and Ecological Network* (Department for Environment, Food and Rural Development 2010).
13 Shannon Hagerman et al, 'Expert Views on Biodiversity Conservation in an Era of Climate Change' (2010) 20 *Global Environmental Change* 192.
14 Nick J.B. Isaac et al, 'Defining and Delivering Resilient Ecological Networks: Nature Conservation in England' (2018) 55 *Journal of Applied Ecology* 2537.
15 Janet Dwyer and Ian Hodge, 'Governance Structures for Social-Ecological Systems: Assessing Institutional Options against a Social Residual Claimant' (2016) 66 *Environmental Science & Policy* 1.
16 William M. Adams, Ian D. Hodge and Lindsey Sandbrook, 'New Spaces for Nature: The Re-Territorialisation of Biodiversity Conservation under Neoliberalism in the UK' (2014) 39 *Transactions of the Institute of British Geographers* 574.
17 See information and examples at: https://www.farmerclusters.com/ (accessed 19 May 2022)
18 Ministry of Economic Affairs, *The Cooperative Approach under the New Dutch Agri-Environment-Climate Scheme* (Ministry of Economic Affairs 2016).
19 Isabella Tree, Wilding. *The Return of Nature to a British Farm* (Picador 2018).
20 Heera Lee and Sven Lautenbach, 'A Quantitative Review of Relationships between Ecosystem Services' (2016) 66 *Ecological Indicators* 340.
21 Defra, 'Defra's Payments for Ecosystem Services Pilot Projects 2012–2015. Review of Key Findings.' (2016).
22 Law Commission, 'Conservation Covenants', LAW COM No 349 (2014).
23 Tom H. Oliver et al, 'Declining Resilience of Ecosystem Functions under Biodiversity Loss' (2015) 6 *Nature Communications.*
24 Martin Scheffer et al, 'Catastrophic Shifts in Ecosystems' (2001) 413 *Nature* 591.
25 Martin J. Westgate, Gene E. Likens and David B. Lindenmayer, 'Adaptive Management of Biological Systems: A Review' (2013) 158 *Biological Conservation* 128.
26 Carl Folke et al, 'Adaptive Governance of Social-Ecological Systems' (2005) 30 *Annual Review of Environment and Resources* 441.
27 Timothy Karpouzoglou, Art Dewulf and Julian Clark, 'Advancing Adaptive Governance of Social-Ecological Systems through Theoretical Multiplicity' (2016) 57 *Environmental Science & Policy* 1.
28 Per Olsson, Carl Folke and Fikret Berkes, 'Adaptive Co-Management for Building Resilience in Social-Ecological Systems' (2004) 34 *Environmental Management* 75.

29 Judith Westerink et al, 'Landscape Services as Boundary Concept in Landscape Governance: Building Social Capital in Collaboration and Adapting the Landscape' (2017) 60 *Land Use Policy* 408.
30 Mark Lubell, 'Collaborative Partnerships in Complex Institutional Systems' (2015) 12 *Current Opinion in Environmental Sustainability* 41.
31 Ian Hodge and William M. Adams, 'Short-Term Projects versus Adaptive Governance: Conflicting Demands in the Management of Ecological Restoration' (2016) 5 *Land* 39.
32 Speech to WWF, 21 July 2017, https://www.gov.uk/government/speeches/the-unfrozen-moment-delivering-a-green-brexit (accessed 19 May 2022).

8

JUDICIAL REVIEW AND NATURE CASES IN THE UK

Current challenges and future opportunities

Carol Day[1]

Introduction

We are all acutely aware of the shocking declines in biodiversity at both UK and global scales. Over half of UK species showed negative trends between 1970 and 2016, with some 41% showing strong or moderate decreases in abundance; and of the 8,431 species assessed using IUCN Regional Red List criteria, 15% are threatened with extinction from Great Britain, and 2% are already extinct.[2] A new measure assessing the intactness of a country's biodiversity suggests that the UK has lost significantly more nature over the long term than the global average and suggests that the UK is among the most nature-depleted countries in the world.[3]

These losses reflect the threats facing the fabric of the planet. In 2009, then director of the Stockholm Resilience Centre ('SRC'), Johan Rockström, and 28 internationally renowned scientists, identified nine processes that regulate the stability and resilience of the Earth system. They proposed quantitative 'planetary boundaries', the crossing of which increases the risk of generating large-scale abrupt or irreversible environmental changes. In 2015, SRC research concluded that four of the nine planetary boundaries had been crossed as a result of human activity[4] including climate change, loss of biosphere integrity (biodiversity loss and extinctions), land-system change and altered biogeochemical cycles (phosphorus and nitrogen).

Stemming and addressing this torrent of loss requires action on all fronts. While environmental NGOs continue to campaign, citizen action in the form of direct action is enjoying a renaissance. Individuals and NGOs are also increasingly turning to the law as crowd funding provides a platform to support legal action for those concerned about climate change, biodiversity loss and public sector intransigence. Such cases are often symbolic and strategically significant.

DOI: 10.4324/9780429299100-8

While the success rates associated with Judicial Review (JR; the mechanism used in the UK to challenge the decisions, acts and inaction of public bodies) are low,[5] the mechanism remains a vital backstop for checking the abuse of public power – and never has that power been more precarious and prevalent than now.

The rule of law and the judicial system

The rule of the law is a principle of the (uncodified) UK constitution that means politicians and public officers at all levels must govern within their powers, that the law applies equally to all and that the law must be accessible, intelligible, clear and predictable.[6] It is the foundation of any democracy. In the Supreme Court case of *Alconbury*,[7] Lord Hoffman described the significance of Judicial Review in the following terms: '*The principles of Judicial Review give effect to the Rule of Law. They ensure that administrative decisions will be taken rationally in accordance with a fair procedure and within the powers conferred by Parliament*'.

Given that JR also represents the final mechanism for civil society to challenge unlawful decisions by public bodies affecting the environment in the courts, it is imperative that it operates effectively. However, for some years, the UK has been in breach of the access to justice provisions of the UNECE Aarhus Convention[8] with respect to the prohibitively high cost of legal action.[9] The UK is also facing criticisms that the scrutiny currently exerted by the Courts in JR is insufficiently intense to comply with the Convention.[10] It could be argued that while JR remains an inherently risky, narrow and expensive process for claimants, why pursue it? But that is to misunderstand the strategically important role that JR can play in environmental protection. I return to this issue at the end of the chapter, after reviewing some of the ways in which JR, and other participatory rights, need to be strengthened in primary legislation.

The intensity of Judicial Review

From the claimant's perspective, a central weakness of JR is that it is rarely concerned with the 'merits' of a decision, or whether the public body has made the 'right' decision – the only question before the court is whether the public body has acted unlawfully in accordance with established legal principles. In particular, it is not the task of the courts to substitute its judgment for that of the decision maker, although it can intervene in appropriate circumstances.

The main grounds for JR (which are neither exhaustive nor mutually exclusive)[11] include: (1) illegality (primarily in not applying the correct statutory test(s)); (2) irrationality (*Wednesbury* unreasonableness); (3) procedural unfairness; and (4) incompatibility with the European Convention on Human Rights and/or EU law. The process of JR contrasts markedly with the function of a planning appeal. Developers challenging a Local Planning Authority's refusal of planning permission (or the conditions attached to a Decision Notice) can apply for a full merits review within six months of the Decision Notice being issued.

Third parties (for whom there is no right of appeal)[12] must apply to the Court for a JR of the lawfulness of a decision within six weeks.

The only review of the 'merits' of a decision that can currently take place in JR is to consider whether the decision was '*Wednesbury* unreasonable'. The *Wednesbury* test was later articulated in *Council of Civil Service Unions v Minister for the Civil Service*[13] by Lord Diplock as a decision: '*So outrageous in its defiance of logic or accepted moral standards that no sensible person who had applied his mind to the question to be decided could have arrived at it'*. It essentially means the court does not intervene and set aside an administrative decision unless it is so outrageous as to be perverse.[14]

Wednesbury unreasonableness is a very high threshold to reach – and there is no special provision for environmental cases. The consequence of this limitation is that the challenges pursued tend to rely on procedural grounds. This can render JR ineffective, as the decision-maker can simply make the same decision again with any procedural irregularities rectified. For example, in *R (Andrew Cawdron) v North Norfolk Council and Balfour Beatty Civils Ltd*,[15] the Council agreed to quash approval for the North Norfolk Distributor Road within weeks of legal proceedings being issued. While local residents were initially delighted, it soon became apparent that the Council intended to remit the decision back to the Planning Committee with the acknowledged procedural defect rectified on the papers. Planning permission was duly granted again within a matter of weeks, essentially rendering the JR procedure, in the eyes of the residents, little more than a costly and time-consuming exercise in delay.

The judiciary has been asked to consider whether *Wednesbury* is the appropriate standard of review in numerous environmental cases in recent years, including those concerning the Habitats Directive.[16] In the case of *Foster*,[17] the English High Court addressed an assessment process under Article 6 of the Directive. The claimants sought permission to apply for a JR of the decision of the Forest of Dean District Council to grant planning permission for a mixed-use development in Gloucester. Their objections were based on the potential impact on bats (in particular, the lesser horseshoe bat) and bat roosts nearby. Counsel for the claimants referred to the judgment of the Court of Justice of the European Union (CJEU) in *Sweetman I*,[18] submitting that a *Wednesbury* standard of review would not reflect European law and that it was clearly for the national court to establish whether the assessment of the implications for the Special Area of Conservation ('SAC') met the requirements of the Directive. The judge (Cranston, J) observed there was '*an air of unreality about this submission*', in that the CJEU could not have been suggesting that national courts must decide when the assessment has *lacunae* (i.e. whether it contains complete, precise and definitive findings) and whether its conclusions are capable of removing all reasonable scientific doubt as to the effects of the works proposed on the protected site concerned. Mr Justice Cranston observed that this approach was also to misunderstand the role of courts in European societies and that, to his mind, the CJEU was simply stating that the national court had to evaluate the assessment in the ordinary way, not become the primary decision-maker. The judge also stated that counsel's submission was contrary to binding authority (albeit in a different context), referring to the case of *Smyth*.[19] In this case, Sales LJ had rejected a submission that in applying the Habitats Directive the national court must apply a more intensive standard of review, in effect making its own assessment afresh. The view in *Foster* was again confirmed in *Abbotskerswell Parish Council v Teignbridge District Council*.[20]

In the case of *McMorn*,[21] which concerned an application for a license to shoot Buzzards *(Buteo buteo)*, Mr Justice Ouseley temporarily opened up the possibility that *Wednesbury* principles can accommodate a more intensive review and that such a position was consistent with *Smyth*. His view appears to have been based on the fact that *McMorn* engaged substantive rights but, in any event, any departure from the standard approach was short-lived. In *Dillner*,[22] the claimant and a number of environmental campaigners sought to challenge the tree-felling policies of Sheffield City Council. One of the claimants argued that the tree-felling required an Environmental Impact Assessment (EIA) and, thus, the claim fell within the protections conferred by the Aarhus Convention. The claimants argued that the case required intense scrutiny, relying on the judgment of Ouseley, J in *McMorn*. Mr Justice Gilbart robustly rejected the argument, citing *Smyth* and *Evans* to re-state the orthodox position that while the status of a claim as an Aarhus claim may be relevant to the question of cost protection, it makes no difference to the requisite standard of review.

This position was reaffirmed in *R (on the application of Mynydd v Gwynt Ltd) v Secretary of State for Business, Energy and Industrial Strategy*,[23] in which the Court of Appeal set out a number of principles for appropriate assessments under Article 6(3) of the Directive. The final principle was that the relevant standard of review by the court is the *Wednesbury* rationality standard, and not a more intensive standard of review, again citing *Smyth*. See also *R (on the application of the RSPB v Natural England)*,[24] in which Mrs Justice Lang DBE held that Natural England had acted lawfully in granting a licence for a trial concerning the brood management of Hen Harriers (*Circus cyaneus*) under the Wildlife and Countryside Act 1981.

In addition to the cases cited above, the limitations of the *Wednesbury* test are borne out by personal experience as a practising lawyer. In 2013, Leigh Day and Landmark Chambers established the Environmental Planning and Litigation Service ('EPLS') to provide specialist advice on prospective environmental JRs. Between 2013 and 2018,[25] the initiative advised some 140 clients as to whether they had arguable grounds to challenge the decisions of public bodies by way of JR. Of those enquiries (most of which concerned planning proposals), counsel advised that 15 cases had grounds for JR with reasonable prospects of success (i.e. they demonstrated identifiable legal errors extending beyond merits review type complaints). Other cases raised substantive review issues, but clients were advised they would not meet the threshold for *Wednesbury* review, despite instances where, for example, decisions had been made on the basis of very little information, or it was clear that only scant consideration had been given to consultation comments from the public.

Most individuals and community groups understandably have limited experience with the process of JR. Many allege the decision-maker has acted unreasonably and unlawfully, only to be advised that the decision cannot be challenged because the courts will not intervene in matters of judgement or on the merits. Most go away disappointed and baffled that the law does not allow them to question a decision that seems indefensible as a matter of common sense, or ask for a review on the basis of what is reasonable or proportionate or better informed. Many point out the inequity that exists between third parties and developers, the latter enjoying the right to appeal a decision and receive a full merits review.

The intensity of review applied in JR is partly a function of the degree to which the courts consider it necessary to defer to the executive. Where the decision-maker has discretion to balance competing considerations, the courts tend to be even more deferential. Thus, in the majority of town and country planning cases, the view of the court is that it is entirely for the decision-maker to attribute to the relevant considerations such weight as it thinks fit. In fact, the courts can, and do, conduct a forensic analysis of the relevant technical issues in some cases (for example in private environmental law cases). The point is that they choose not to do so in public law cases.

The Aarhus Convention requires contracting parties to ensure that members of the public concerned have access to a legal review procedure to challenge the substantive and procedural legality of decision, acts or omissions subject to Article 6 of the Convention (essentially cases concerning EIA or other processes subject to public consultation) and other environmental cases. Unfortunately, the Convention does not define 'substantive legality' and there are varying standards of review available in the courts and tribunals of the parties to the Convention, some of which (as in the Swedish Land and Environment Courts) provide a full merits review. The CJEU applies a proportionality test in environmental cases.[26] The intensity with which it is employed varies depending on whether the national measure interferes with a freedom guaranteed by an EU treaty, relies on derogation from an EU treaty, or simply implements EU law.[27]

The Aarhus Convention Compliance Committee (the 'Compliance Committee') has previously suggested that the application of the proportionality principle could provide a more appropriate standard of review in cases within the scope of the Convention in the UK, provided that the principle does not exclude any issues of substantive legality from review.[28] In 2017, the RSPB, Friends of the Earth, Friends of the Earth Scotland and Leigh Day submitted a Communication to the Compliance Committee ('C156') alleging that the UK fails to comply with the requirement to provide a review of procedural and substantive legality.[29] The Communication was declared admissible in March 2018, a hearing was held in November 2019 and, following subsequent written questions posed by the Committee, proceedings remain ongoing as at September 2022.

In the interim, there was a notable further judgment of the English High Court in May 2019. The Heathrow case[30] encompassed five JR challenges to the Secretary of State for Transport's decision to designate an Airports National

Policy Statement (ANPS). The ANPS sets out the Government's policy on the need for new airport capacity in the South East of England and its preferred location – namely a new runway at Heathrow. Four of the claims raised a combination of 22 environmental grounds encompassing climate change, air quality, habitats, SEA, surface access and consultation. In a lengthy and detailed judgment, the Judges dismissed all 22 grounds, most of which (despite a ten-day hearing) were refused permission to proceed.

The appropriate standard of review to be applied in this case was a matter of considerable debate during the hearing. While the claimants argued that the nature of the issues involved warranted a high level of scrutiny, the court was of the view that a 'low intensity' of review is appropriate in cases involving political judgement, where the decisions relate to a matter of national economic policy and confirmed that it would not intervene outside of the extremes of bad faith, improper motive or manifest absurdity. The judges reiterated two, by now established, principles. First, in cases where the grounds involve matters of scientific, technical and predictive assessment, the courts afford an enhanced margin of appreciation to a decision-maker (*Mott*[31]). Second, cases concerning Articles 6(3) and (4) of the Habitats Directive do not involve any derogation from fundamental freedoms or rights of the kind with which the principles set out in *Lumsdon* were concerned.[32] As such, the appropriate standard of review is the *Wednesbury* rationality standard (again citing *Smyth*[33]). As it is hard to imagine a case raising issues of higher public interest, this judgment compounds concerns that the courts have almost entirely set their minds against a more intense standard of review in Aarhus JR claims.

On appeal, the Court of Appeal upheld the decision of the High Court on this issue: 'there is no justification for applying a more intense standard of review than "Wednesbury" to the operation of the provisions of article 6(4) of the Habitats Directive,' held the Court.[34] 'Neither the court's decision in [Case C-723/17] *Craeynest* nor the Advocate General's opinion supports a different conclusion.'[35] Similarly, the Supreme Court later agreed that *Wednesbury* was the appropriate standard of review.[36]

The Court of Appeal revisited the standard of review in the High Speed 2 ('HS2') case challenging the Government's decision to proceed with Phase 1 of the HS2 project between London and Birmingham in 2020.[37] The £80.7bn project (now thought to be nearer to £106bn) directly affects 32 ancient woods and emissions from construction of the full HS2 network are estimated to be between 8 m and 14 m tonnes of CO_2e (carbon dioxide equivalent) over the construction period, such that the project will not become carbon neutral for at least 120 years. A challenge to the Cabinet decision to proceed with Phase 1 was brought by the naturalist and campaigner Mr Christopher Packham CBE. Following a refusal of permission for Judicial Review in the High Court in April 2020, a 'rolled-up' hearing took place in the Court of Appeal in July 2020 and judgment was handed down in July 2020. The Court of Appeal discussed the appropriate intensity of review to be applied in the case, holding that: 'In our view, however, this is unquestionably the kind of case in which the court should refrain from anything beyond a "light touch" approach, applying the traditional test of "irrationality"'.

If the Compliance Committee agrees with the Communicants in C156,[38] the UK should be prompted to decide what intensity of review is Aarhus-compliant and how it might be guaranteed. In its comments on the UK's Response to the Communication, the Communicants accepted that the *Wednesbury* test has *the potential* to act as an effective form of substantive review consistent with Article 9(2) or 9(3) of the Convention. The problem is that it is almost universally understood and applied narrowly. One mechanism for ensuring that it is applied consistently, as required by Article 3(1) of the Convention (which imposes an obligation on parties to establish and maintain a clear, consistent and transparent legal framework), would be to pass legislation requiring *Wednesbury* to be operated and applied in a more intensive way, although there may of course be other ways in which the UK could ensure a compliant and effective standard of review.

Pending the outcome of C156, the Compliance Committee recently made findings against the UK in a separate case raising related issues.[39] In that case, the Committee held that by the High Court of Northern Ireland not undertaking its own assessment, based on all the evidence before it, of whether:

i The development in question was 'likely to have significant effects on the environment by virtue of factors such as its nature, size or location';
ii The permit conditions could be implemented in practice without adverse environmental impacts,

but instead relying on the assessment of the public authority that took the contested decisions, the UK failed to provide for a review of the substantive legality of those decisions in accordance with the requirements of Article 9(2) of the Aarhus Convention. In key passages, the Compliance Committee found that:

> 119. [...] article 9(2) [of the Aarhus Convention] requires that a reviewing body must review all the facts, evidence and arguments before it and, based on that review, determine whether the contested decision is lawful. This requires the court to carry out its own assessment, in the light of all the evidence before it, as to whether the applicable legal requirements were met. The court must also clearly set out its reasoning when doing so.
>
> 120. For example, in a challenge to the substantive legality of an EIA screening decision, the court must make its own assessment, based on all the evidence put before it, as to whether the proposed activity was likely to have a significant effect on the environment and thus to require an EIA. It would not be sufficient to merely check that the decision-maker carried out the correct procedural steps for determining whether a project was likely to have significant effects. Nor does it suffice for the court to check that the decision-maker had formally applied the correct legal test and that the decision-maker had convinced itself that that test was met in a particular case.

> 121. To be clear, the Convention does not require the court to undertake a completely fresh analysis of all matters arising in the case and to substitute its decision for the decision taken by the competent authority. Nevertheless, the court must undertake its own assessment of all the evidence before it to determine whether the applicable legal requirements were met. The Committee considers that this requires the court to perform a review function over findings of fact and the weight to be given to evidence where those may have a direct impact on the determination as to whether the applicable legal test (for example, likely significant effects) has been met.

This is precisely the sort of approach that was rejected by the High Court in *Foster*[40] in 2015 as having an 'air of unreality' about it, and it is in stark contrast to the orthodox approach of the UK courts. How the UK government will address these findings remains to be seen.

The cost of taking legal action

The cost of taking legal action in the UK has long been subject to the sarcastic aphorism, that 'in England justice is open to all – just like the Ritz Hotel'.[41] My own first foray into JR (in fact in Scotland) was an unsuccessful challenge to the construction of a funicular railway up Cairngorm mountain in Aviemore, resulting in an adverse costs order against WWF of £195,500 plus VAT.[42]

In 2005, the high costs of pursuing legal action in the UK prompted a coalition of environmental NGOs[43] to submit a complaint to the European Commission arguing that the UK failed to comply with the Public Participation Directive.[44] The Commission pursued infringement proceedings against the UK and some eight years later, the CJEU ruled in favour of the Commission.[45] Interestingly, while the issue of legal costs is addressed very differently by EU Member States, the CJEU position is generous to unsuccessful applicants before the CJEU, where applicants are not expected to pay EU institutions' legal costs on the basis that the EU institutions are already funded by the public purse (in effect, double recovery).[46] The only costs typically payable in practice are the travel and subsistence costs of EU legal staff attending the hearing, which usually (in my experience) amount to no more than a few hundred euros. It is, however, easier for a court to be generous to unsuccessful applicants regarding their costs liability when it exercises a very restrictive approach to standing.

Prior to the CJEU ruling and, the result of a parallel Communication on the same issue submitted to the Compliance Committee,[47] the devolved administrations of the UK introduced bespoke costs rules for environmental (Aarhus) cases. The new rules were a significant improvement on the previous regime of Protected Costs Orders (following the case of *Corner House*[48]) because they offered prospective claimants absolute advance certainty as to the extent of their adverse costs liability. The level of the cap was set at £5,000 for individuals and £10,000 in all other cases.

While the cap on adverse costs is helpful, it should be noted that the Civil Procedure Rules (CPR) also limit the amount of legal costs that successful claimants can recover from the defendant public body to £35,000 inclusive of VAT (this is called the 'reciprocal cap' or the 'cross cap'). The effect of the reciprocal cap can make cases too expensive to win. Reciprocal caps have no legal basis in the Aarhus Convention. In Communication C33, the UK argued that:

> it is important to recognize that the provision of a fair and just system of law involves treating all parties to litigation fairly. The resources applied by public authorities in defending judicial review proceedings stem ultimately from the taxpayer, and it is therefore proper that the cost implications for both parties in an individual case should be taken into account.[49]

This reasoning flowed through into the 2013 Environmental Costs Protection Regime, which introduced the concept of the reciprocal cap. However, this type of argument fails to recognise that the executive has very substantial sums of public money available to it, whereas claimants may not. The Compliance Committee takes a holistic approach to the issue of costs,[50] recognising that the costs of the proceedings include the applicant's own costs and the effect of reciprocal caps. The Compliance Committee has previously recognised the limiting effect of reciprocal caps[51] and has also already clarified that in the context of Article 9(4) of the Convention, 'fairness' refers to what is fair for the *claimant*, not the defendant public body.[52]

In 2017, the Ministry of Justice (MoJ) implemented a number of changes to the costs regime for Aarhus cases in England and Wales. These changes were progressed in the face of substantial public and Parliamentary opposition[53] and irrespective of the fact that environmental cases constitute less than 5% of JRs taken (approx. 120 of some 3,600 cases annually).[54]

The most significant change was that the new Rules removed prior certainty for claimants with regard to adverse costs liability in environmental cases.[55] When making an application for JR, claimants are now required to provide the court with a schedule of financial resources covering their significant income, expenditure, assets and liabilities. On the basis of this information, defendants may apply to the Court for the 'default caps' of £5,000 and £10,000 to be varied. The Government's clear expectation, as set out in the preceding public consultation,[56] was that the caps would universally be increased – as has indeed proven to be the case (see below).

While at that time the court had to ultimately ensure that costs would not be 'prohibitively expensive' for the claimant as a result of EU law,[57] many feared the prospect of an increased cap at some unknown point in the future would have a 'chilling' effect on potential claimants. This fear was immediately materialised when the Liverpool Green Party (LGP, an unincorporated association, which can only bring a claim through an individual who acts on behalf of its members) sought to challenge Liverpool City Council's decision to grant planning permission for a car park in an Air Quality Management Area without undertaking an air quality assessment. In response to the LGP's Pre-Action Protocol letter, the Council failed to properly engage with the substance of the grounds, but noted that:

> ... any claim for a cost protection order will be carefully examined. In particular it is noted that the court now has discretion under CPR 45.44 to vary the limits on maximum costs liability for Aarhus Claims and the Council will therefore require confirmation of the financial resources of your client in the event that it seeks a protective costs order.

In light of this correspondence, LGP was unable to find an individual prepared to act as claimant.

Numerous other aspects of the current Environmental Costs Protection Regime ('ECPR') in England and Wales remain problematic for those seeking to bring environmental cases. Some environmental claims (including private law claims such as private nuisance within the scope of Article 9(3) of the Convention) are not covered by the ECPR – a position found to be incompatible with the requirements of Article 9(4) of the Convention by the Compliance Committee in 2015.[58]

At the time of writing, and as predicted, any variation of the costs cap has almost universally been upwards. The Compliance Committee has previously observed that the levels of the default costs caps can only be acceptable if variation downwards is not only theoretically available but can be predictably relied upon in practice and that the relatively high proportion of cases in which defendants sought an increase in the costs cap may create a deterrent effect.[59] This contrasts with the approach in Northern Ireland, in which default caps can only be varied downwards and the reciprocal cap can only be varied upwards, which represents a significant improvement on the current regime in England and Wales.

Moreover, in the case of *R (Bertoncini) v London Borough of Hammersmith and Fulham and Kendall Massey*,[60] the judge held that an Interested Party ('IP') has standing to apply for a variation of the default cap. In this case, the judge maintained a modified default cap of £20,000 and ordered the claimant to pay costs totalling £16,991 (of which £12,000 were to be paid to the IP).

The requirement to provide a schedule of financial resources when applying for JR can have a deterrent effect on potential claimants because their financial circumstances will be provided to the defendant and may be discussed in open court. It is also a practical burden that makes filing a claim harder, and in certain cases can prove to be entirely unnecessary. This includes situations in which a defendant has indicated it will not so apply or where the objective factors all point to a variation being prohibitively expensive due to the importance of the case for the environment and there being no personal interest in its outcome.

Another modification of the 2017 regime was to require separate costs caps for each claimant, which has the practical effect of largely preventing individuals and/or environmental NGOs bringing joint claims.[61] The Compliance Committee sees no basis for this rule, particularly where claimants make the same legal arguments on the same factual basis, and also recognises the benefit of claimants sharing the costs burden for challenges within scope of the Convention. A further unhelpful modification is the basis on which defendants are required to pay the costs of unsuccessfully challenging the status of an Aarhus claim. Prior to February 2017, defendants who unsuccessfully challenged that a claim was an Aarhus claim were required to pay 'indemnity costs' to claimants regarding that challenge. Since February 2017, defendants are required to pay the claimants' costs regarding the challenge on the 'standard' basis, which is lower.

The position regarding costs on appeal also remains unclear. Civil Procedure Rule 52.19A requires the court to consider whether the costs of the proceedings will be prohibitively expensive for a claimant when appealing a decision in an Aarhus claim, requiring it to make an order limiting the recoverable costs where it is considered they will be. However, the current Rule fails to ensure sufficient clarity or costs protection for claimants in appeals – and fails to clarify that the 'not prohibitively expensive' requirement applies to the proceedings as a whole, encompassing all stages of the procedure.

The Compliance Committee has also noted that the 2017 CPR amendments do not provide clarity to applicants seeking interim injunctions as to: (a) whether a cross-undertaking will be required, and (b) if a cross-undertaking is required, what its level will be. Additionally, individuals and eNGOs who join proceedings as interveners in support of the claimant are not currently entitled to benefit from the Convention's requirement that proceedings must not be prohibitively expensive.

The above difficulties are compounded by the level of the court fees. The Administrative Court fee has doubled in recent years to just under £1,000 and fees in the Supreme Court routinely amount to approximately £5,000. Secondly, while public funding (legal aid) is theoretically still available in environmental cases, there are strict merits and means requirements (NGOs do not qualify in any event). Where awarded, a Community Contribution in the order of several thousands of pounds is usually required. As a result, public funding is now rarely pursued in environmental cases.

The impact of these changes has been further compounded by unhelpful jurisprudence. In *CPRE Kent v Secretary of State for Communities and Local Government,* the Supreme Court clarified the costs position for unsuccessful claimants in JRs and statutory reviews involving multiple defendants. When permission is refused, a claimant may now be liable to pay the costs of more than one defendant and/or Interested Party ('IP') to prepare and file an Acknowledgement of Service ('AoS') and Summary Grounds of Resistance ('SGoR'). It is not necessary to show 'exceptional' or 'special' circumstances apply, although costs must be reasonable and proportionate. The effect of this judgment was immediate, with an IP in one case I was involved in immediately seeking costs of just under £24,000 for simply preparing the AoS and SGoR '*per CPRE Kent v Secretary of*

State for Communities and Local Government [2021] UKSC 36'. Permission was subsequently refused. Fortunately, the judge fixed the costs cap at £10,000, ordering the costs of preparing the AoS be paid by the claimant to the defendant in the sum of £8,900.00 and by the claimant to the IP in the sum of £1,100.00.

The immediate effect of the Supreme Court judgment in *CPRE Kent* was two-fold. First, IPs appear to be emboldened to submit excessive estimates of costs at an early stage in the proceedings and, second, the court will, in such cases, now routinely order costs up to the full level of the default Aarhus caps at the permission stage (and that's assuming that the default costs caps are not varied upwards). This position is exacerbated by the position following *Bertoncini* in that IPs can now not only request their costs at the permission stage, but they can also apply for the cap to be varied upwards to accommodate excessive costs estimates.

The Aarhus caps are intended to limit the level of adverse costs exposure throughout the duration of a case to ensure the proceedings are not prohibitively expensive for the claimant. The possibility that sums well in excess of the default Aarhus cap can be ordered at an early stage will have a 'chilling effect' on potential claimants and therefore further undermine the UK's ability to comply with the Convention.

In 2021, the Seventh Meeting of the Parties to the Aarhus Convention adopted Decisions in respect of numerous parties in continuing non-compliance with the Convention. In respect of the UK, Decision VII/8S requests that the UK (*inter alia*) ensures that the allocation of costs in all court procedures subject to Article 9, including private nuisance claims, is fair and equitable and not prohibitively expensive and submit a plan of action, including a time schedule, to the Committee by 1 July 2022 regarding the implementation of the recommendations in Decision VII/8S. The Action Plan was published on 1 July 2022 and has been heavily criticised by eNGOs. Its preparation involved no substantive consultation with eNGOs in England and Wales, and in relation to costs, the plan does not include any actual measures, but instead simply signposts a future consultation regarding the ECPR.

Remedies in JR

The most common remedy in JR proceedings is the quashing order, which nullifies a decision made by a public body. It is usually made where an authority has acted *ultra vires*. Perhaps the second is a declaratory judgment, which clarifies the respective rights and obligations of the parties to the proceedings, without actually making any order.

An injunction prevents a public body from acting in an unlawful way. Where there is an imminent risk of damage or loss, and other remedies would not be sufficient, the court may grant an interim injunction to protect the position of the parties before a full hearing. In this situation, it is possible that the party benefitting from the injunction will be asked to give the court an undertaking that if the other side is successful at the final hearing, the party which had the benefit of the interim protection can compensate the other party for its losses (this is sometimes called a cross-undertaking in damages). For example, in the (now very old) *Lappel Bank*[62] case, the RSPB requested an interim injunction to protect Lappel Bank (part of the Medway Estuary and Marsh system important for two species listed in Annex I of Wild Birds Directive) from development as a port for the duration of the legal proceedings. However, the RSPB was unable to give the court an undertaking that it compensate the Port of Sheerness for losses resulting from the imposition of interim relief. The RSPB subsequently won the case, which went all the way to the House of Lords (as was) and involved a referral to the European Court of Justice (as was) but it was a somewhat pyrrhic victory as, by that stage, the mudflats had been developed as a car park.

The discretionary nature of the remedies outlined above means that even if a court finds a public body has acted wrongly, it does not *have* to grant any remedy. Contrast this with the situation before the CJEU, in which the remedies are genuinely dissuasive. Fines for non-compliance with CJEU judgments routinely extend to an onerous lump sum payment and daily penalties. For example, a failure to comply with judgments concerning the implementation of the Urban Waste Water Directive in Belgium resulted in 2013 in a lump sum fine of €10 million and a penalty payment of €860,000 for each six-month period of delay in taking the measures necessary to comply with the judgment.[63]

While there appear to be few applications for interim relief before the UK courts, the Government sought to restrict the availability of remedies in JR in 2021. Clause 1 of the Judicial Review and Courts Bill directed the courts towards granting two new remedies – Suspended Quashing Orders ('SQOs'), which would only take effect at a certain point in the future, and Prospective Quashing Orders ('PQOs'), which appeared to only have a forward-looking effect. Following opposition from a wide spectrum of civil society organisations and the House of Lords due to the profound implications for the rule of law, the House of Lords (on an amendment from Lord Anderson) removed the presumption that these new remedies would be awarded in certain situations. It is thought the significant opposition demonstrated in the upper house persuaded the Government against reversing the position in the House of Commons prior to Royal Assent in 2022.

The JR climate

Against the limitations outlined above, there have been a number of other unhelpful changes to the JR process in recent years.

Following changes to the Aarhus costs regime in 2013, there is no oral renewal (i.e. no hearing to allow an earlier refusal of permission to be reconsidered) for claims deemed by the permission judge to be "totally without merit" (this applied to some 15% of JR applications in 2021).[64] Second, section 84 of the Criminal Justice and Courts Act 2015 lowered the threshold in the 'Significant difference test', in which the court can refuse an application for JR where it considers that the outcome for the applicant would not have been substantially different if the conduct complained of had not occurred. In a case concerning the lawfulness of a badger cull in Dorset (*Langton II*[65]), the claimant alleged that Natural England's ('NE') approach to assessment in respect of some 45 Sites of Special Scientific Interest ('SSSIs') was fundamentally flawed as it had omitted certain species from consideration when deciding on the licence conditions. NE conceded that it may have erred as regards three SSSIs, but rejected the claimant's case in relation to the other 42 sites. In 25 of these sites, NE contended that the alleged omissions would not make any difference to the conditions which it decided to impose on the licences; thus the claimant's case in relation to these sites was purely academic. The judge agreed, partly on the basis that taking academic points in one case may encourage others to bring similar claims in other cases when the court is already hard-pressed with its caseload.[66]

There is also now a reduced (and very demanding) time limit for challenging decisions under the Planning Acts of six weeks[67] In reality, it is extremely difficult for individuals and community groups to find lawyers, secure legal advice, fundraise and decide to embark on legal action within that deadline.

Brexit, governance and Judicial Review

The EU referendum (UK Brexit vote) on 23 June 2016 resulted in 51.9% of the votes cast being in favour of leaving the EU. Although the referendum was legally non-binding, the government of the time promised to implement the result.

Pre-Brexit any individual, community group or NGO in the UK could submit a complaint to the European Commission concerning an alleged breach of EU law. The Commission would investigate the complaint and has the discretion to pursue it without the complainant having any financial exposure or obligation to resource it. The Commission would attempt to resolve the issue with the Member State through a pre-litigation procedure involving the issuing of a letter of formal notice and a reasoned opinion, both of which give the Member State a fixed time to comply. If these measures failed, the Commission could then refer cases to the CJEU, which would reach a judgment and, where necessary, impose fines (further to a request from the Commission) to ensure compliance.

This crucial enforcement role enabled individuals, community groups and environmental NGOs to secure significant environmental safeguards. For example, in 2012, WWF-UK complained that the UK had failed to designate Special Areas of Conservation (SACs) for the harbour porpoise *(Phocoena phocoena)*, a species listed in Annex II of the Habitats Directive. The complaint resulted in a judgment against the UK[68] and the subsequent designation of seven large marine SACs for this species. The complaint concerning the high cost of legal action, which resulted in the introduction of Aarhus costs regimes across the UK (as discussed above), is another example.

In 2017, Michael Gove, in his first keynote speech as Environment Secretary, outlined the government's ambitious 'Green Brexit' vision to outclass the

EU's environmental regime. The UK/EU Trade and Cooperation Agreement ('TCA') includes a non-regression clause[69] designed to prevent the UK and EU softening environmental protections. The Government has repeatedly stated that it has no intention of doing so (an issue to which I return below).

Following delays caused by a general election and the Covid-19 pandemic, the Environment Act 2021 ('the Act') received Royal Assent on 9 November 2021, nearly three years after a Bill was first proposed to govern environmental matters after the UK's departure from the EU.

The Act establishes a new environmental governance system for England which comprises four interdependent elements. First, the preparation of an environmental improvement plan to chart and drive government's progress on the environment.[70] Second, it gives the Secretary of State the power to set legally binding, longer-term, environmental targets lasting at least 15 years, across a variety of areas relating to the environment. More specifically, the Government must set at least one environmental target in each of four priority areas (water, air quality, biodiversity and waste/resource efficiency), as well as on species abundance and fine particulate matter (PM2.5). Third, the Secretary of State is required to prepare an environmental principles policy statement to embed environmental principles at the heart of policy development across government.[71] Finally, chapter two of the Act concerns the creation of a new body, the Office for Environmental Protection ('OEP'), to replace the EU's former role in holding the UK to account on environmental law. There have been rapid developments in all of these areas following Royal Assent for the Act, some of which are considered below.

The Office for Environmental Protection

While the establishment of the OEP aligns with the Rule of Law requirement for effective implementation of the law, the new watchdog has been described as 'impartial' rather than properly 'independent'.[72] The Secretary of State can issue guidance to the OEP on its enforcement policy and on its enforcement functions, and the OEP must have regard to this guidance. The Secretary of State is closely involved in appointing, paying and reappointing the members of the OEP. It is thought that these factors will limit the extent of the OEP's independence.

The Act also establishes a new court procedure for the OEP when pursuing breaches of environmental law by a public authority – Environmental Review ('ER'). The default outcome in an ER where there has been a breach of environmental law is the 'statement of non-compliance'.[73] Crucially this statement does not affect the validity of the unlawful conduct (thus reversing the current position that unlawful acts are presumed to be void, but with judicial discretion to validate them). This new default undermines the principle of legality with a presumption that unlawful acts by a public authority are valid.

Further remedies are available under the Act, but they suffer significant restrictions. A court in an ER is not permitted to quash an unlawful action by a public authority if doing so risks causing substantial hardship to, or would substantially prejudice, a third party.[74] This means that a clear unlawful action must be allowed to continue if a third party stands to benefit from it – or would suffer a loss if it does not continue. This approach fails to take account of other factors such as harm to nature and to people. The Act provides that damages cannot be granted following an environmental review. There is nothing wrong in principle with prohibiting a court from awarding damages to the OEP. However, in the absence of any ability for the OEP to issue fines for a breach, this restriction tends to further weaken the effective implementation of environmental law. These provisions on the remedies available in an environmental review undermine the rule of law.[75] More generally, at the time of writing it is unclear how the parallel processes of Environmental Review and Judicial Review will relate to one another. Will the OEP continue to investigate a complaint that is the subject of a JR before the courts (and vice versa)? Much remains to be seen.

In January 2022, the OEP invited views on its draft strategy and enforcement policy.[76] Both Greener UK and Wildlife and Countryside Link emphasised the importance of the OEP acting objectively and independently from Government, taking a proactive approach to enforcement and establishing a clear approach to working with other public authorities such as the Climate Change Committee and other arms length bodies. Both groups also urged the OEP to ensure the OEP's approach to Northern Ireland issues is firmly and holistically embedded in the final strategy and enforcement policy, following the agreement of the Northern Ireland Assembly to commence the relevant parts of the Environment Act.

Environmental targets

In March 2022, Defra launched an eight-week public consultation on the new, long-term environmental targets required under the Environment Act, including for air quality, water, biodiversity and resource efficiency/waste reduction.[77] In responding to the proposals, environmental NGOs were quick to point out that the Government's 2018 25-year plan[78] promised to leave the environment in a better state for the next generation.[79] However, the consultation paper predicts that species abundance will continue to decline until the 2030 target, followed by a slow recovery, with a target for species abundance in 2042 to be increased by just 10% on 2030 levels. That will mean that at best wildlife levels in 2042 will be back at 2022 levels, and could mean that by 2042 there is actually less biodiversity in England than there is today. It is also concerning that, except for some seabirds, no marine wildlife will be included in species to be targeted and measured under the 2030 goal.

Moreover, there is no overall target for the quality of rivers and streams. Instead, the Government is proposing targets to reduce pollution from particular sectors: agriculture and water. The EU Water Framework Directive set a target for overall ecological condition of rivers, but that target expires in 2027, so there will be no long-term ambition in law for improving water quality. Compliance with targets for water companies and for farmers will be assessed on the basis of modelling data, based on actions taken, rather than real-world assessments of pollution in rivers. This leaves considerable scope for error and for the overall condition of rivers and streams to be obscured. Wildlife and Countryside Link is calling for a new, more ambitious target for the overall condition of rivers, streams and smaller water bodies and the deadline of 2040 to be brought forward to 2030 given the importance to public and environmental health.

Similarly, the 25 Year Environment Plan included a target for 75% of Sites of Special Scientific Interest ('SSSIs') to be in good condition by 2042. This is a non-statutory, non-binding target which has not been included in the proposals. Only 38% of SSSIs are now in favourable condition,[80] and without a strong and legally binding target for England's most important nature sites, the chances of halting nature's decline are significantly undermined.

Environmental Principles Policy Statement

In June 2022, Defra published the Environmental Principles Policy Statement ('EPPS') to embed environmental principles at the heart of policy development across government. Section 19(1) of the Act places a legal duty on ministers of the crown to have 'due regard' to the EPPS in their policy making. The government has also indicated that there will be an implementation period of unknown length to allow government departments to prepare for the duty. Thus, the duty may not be in force for an unknown and potentially considerable length of time.

Paragraph 203 of the explanatory notes to the Environment Act 2021 clarifies that this means that, when making policy, ministers must consider the EPPS with substance, rigour and an open mind. There is no clarity on how the principles will be used to actively inform policy making before and during any implementation period. Meanwhile, ministers in other government departments are actively criticising the environmental principles, taking a particular interest in the precautionary principle. There are also, at the time of writing, proposals to revoke swathes of environmental law that exists in the form of retained EU law, some of which relates to environmental principles.[81]

The EPPS is a key tool to deliver the flagship aim of the government's 25 Year Environment Plan to leave the environment in a better state than it found it. The Act already excludes certain areas of fiscal and defence policy from this process, adding further restrictions via the policy statement itself would severely inhibit its ability to embed environmental principles in all Government policy making and would not be in keeping with the intention of the Act.

There are also concerns about the principles themselves. The Act defines the integration principle in section 17(5) as 'the principle that environmental protection should be integrated into the making of policies', but the revised statement merely advises policy makers to 'look for opportunities to embed environmental protection in fields of policy that have environmental effects'.

Similarly, the EPPS relies on the 1992 Rio Declaration definition of the precautionary principle 'where there are threats of serious or irreversible environmental damage, a lack of full scientific certainty shall not be used as a reason for postponing cost-effective measures to prevent environmental degradation'. This approach has long been criticised for the high threshold for action (serious or irreversible) and the counter-intuitive assumption that in a state of uncertainty the cost-effectiveness of decisions can be calculated.[82] The statement also places the precautionary principle under a 'risk management' heading, suggesting the Government is promoting a US-style approach of preventing projects or substances only when there is demonstrable evidence that proceeding would lead to serious or irreversible harm. The revised statement also frames the precautionary principle as a tool for managing the risk of bringing forward innovative policies that may cause harm to the environment, which will further restrict its application. The precautionary principle should instead be given its wider interpretation in the EPPS, to guide policy makers to take a sensible approach in applying or recommending measures to prevent environmental harm in the absence of full scientific certainty about the severity or cause of the harm.

The Habitats Regulations 2017

Finally, the Act also gives the Secretary of State the power to amend the Conservation of Habitats and Species Regulations 2017 (S.I. 2017/1012).[83] In making such regulations the Secretary of State must have regard to 'the particular importance of furthering the conservation and enhancement of biodiversity' and may do so 'only if satisfied that the regulations do not reduce the level of environmental protection provided by the Habitats Regulations'. In March 2022, Defra published a Nature Recovery Green Paper ('the Green Paper'), with plans to 'refocus' the Habitats Regulations to deliver creative public policy thinking that delivers results, 'rather than relying on change being set principally by litigation and case law'.[84] The Green Paper invited views on the relative merits of various options covering site protection and Habitats Regulation Assessment ('HRA').

Many environmental groups expressed concern about the proposals, pointing out that the current international and domestic legal regimes for biodiversity protection are well-established, are generally working well and provide practitioners with certainty and clarity. Moreover, while high level and lacking in detail, it was clear that the proposals would (if implemented as proposed) involve significant changes to the existing domestic system for the protection of habitats and risk weakening the existing habitats protection regime. Certain proposals appear to have an intentionally deregulatory effect and, if implemented, would compromise the Government's ability to achieve the legally binding targets it has set itself in ss.1–5 of the Environment Act 2021 and other legal instruments.

Proposals for HRA include 'streamlining' and deregulating the consenting process, including a proposal to allow the consideration of mitigation proposals and the mitigation hierarchy at an earlier stage in the assessment process, thus restoring the UK to the legal position which existed prior to Case C-323/17 *People Over Wind and Sweetman*. Similarly, the suggestion that the precautionary principle should in future be applied with greater 'flexibility', indicates a watering down of a fundamental principle of environmental regulation, as is the suggestion that it should be made easier to derogate from the requirement not to adversely affect the integrity of protected sites in the case of planning applications for certain types of climate change infrastructure. Taken as a whole, the Green Paper proposals undermine the rigour of the Habitats Directive regime, which has been the keystone for biodiversity protection in the UK for the last 30 years and could seriously compromise the UK's international commitments by weakening environmental protection.

An Environmental Court or Tribunal ('ECT')

It would not be apt to end this chapter without a reference to the potential benefits of a bespoke environmental court or tribunal.

There has been an exponential growth in such courts or tribunals in recent decades. By 2016, more than 1,200 (not including those at local or municipal level) were dispersed in 44 countries world-wide[85] According to leading experts in the field, Professors Pring and Pring, this development is '*dramatically changing the playing field for environmental justice around the world*'.[86] As of 2018, nearly 1,500 ECTs existed.[87] This expansion has engendered significant attention for the way in which it has changed judicial responses to environmental problems, according to Brian J Preston SC, Chief Judge of the Land and Environment Court in New South Wales, Australia, and one of the world's leading experts on the development of ECTs. In 2014, the Hon. Justice Preston identified the twelve key characteristics that are required for an ECT to operate successfully in practice.[88] Of crucial importance is the requirement that ECTs are independent of Government and impartial, but other factors with immediate resonance for the UK include the fact that such courts have a comprehensive and centralised jurisdiction covering administrative, civil and criminal proceedings – thus ensuring that cases with one or more elements can be dealt with holistically.

Other key factors seem to be that the judiciary is environmentally literate and has access to scientific and technical experts acting as advisers. The Land and Environment Court in New South Wales has a continuing education programme for judges, including seminars and training programmes on current environmental issues and advances in knowledge.[89] Environmental courts and tribunals also facilitate access to justice and develop environmental jurisprudence.[90]

Other than the First-tier Tribunal (general regulatory chamber), which deals with a very limited range of environmental issues,[91] no such court or tribunal exists in the UK. The issue has been repeatedly raised in England and Wales,[92] amidst concerns around the practicality and cost of restructuring the courts and tribunal system to create a specialised environmental forum. On this point, Judge Preston observed that the ability of the court to look at multiple aspects and deal with issues quickly and effectively meant that the court comprehensively pays for itself.

While the establishment of such a forum has thus far gained little traction in England and Wales, the Scottish Government is required to consult on the establishment of an environmental court, no later than six months after new watchdog Environmental Standards Scotland publishes its first strategy. This consultation will take place by Spring 2023, creating an important opportunity for reassessing how environmental disputes are dealt with in Scotland. The Environmental Rights Centre for Scotland ('ERCS') argues that an environmental court could be designed to ensure litigation is affordable and to improve access to justice and that such a body could be given the power to carry out merits review. Moreover, a single ECT could hear multiple legal issues in the same forum, providing administrative costs savings and increasing convenience for the parties and could appoint technical or scientific members to sit alongside judges, thus allowing judges to develop specialist expertise.[93]

A specialist forum could have bespoke rules to ensure compliance with the access to justice provisions of the UNECE Aarhus Convention. For example, rules on time limits for issuing proceedings could ensure fairness to claimants in accordance with Article 9(4) of the Aarhus Convention and confirm the liberal standing requirements for JR should be maintained (and extended to statutory reviews). It could provide a review of procedural and substantive legality as required by the Convention (although many ECTs apply a full merits review[94]) and be staffed by environmentally literate judges assisted by scientific and technical experts.

The creation of a specialist forum could provide the opportunity to address the challenges around prohibitive expense outlined above and consideration could also be given to the requirement that certain environmental cases could have an automatic suspensory effect.[95] The court could enjoy the power to award dissuasive and innovative remedies including the power to reverse a decision of a consent authority or set aside a provision of a plan,[96] restoration orders, compensation for any loss or damage suffered by the claimant (in addition to any order as to legal costs), exemplary damages as appropriate.[97] and the capacity to make individuals personally accountable for the shortcomings of businesses over which they exert control. It could also have the power to impose hefty fines in light of non-compliance with its judgments. Finally, there could be public funding for cases brought in the public interest by individuals and NGOs. People should not have to prove they are 'poor enough' – and communities should not be expected to raise substantial sums of money – to qualify for it. Such a fund could guarantee compliance with Article 9(5) of the Aarhus Convention, which requires contracting parties to consider the establishment of appropriate assistance mechanisms to remove or reduce financial and other barriers to access to justice.

Concluding remarks

During my time working on UK and EU wildlife policy for WWF, the organisation (for it was originally a fund for nature) supported numerous JRs (including many *causes célèbres* such as Oxleas Wood, Twyford Down, the Newbury Bypass and Heathrow). My experience of those cases not only convinced me to become a lawyer, they taught me that losing a legal battle (because, realistically, we lose most of them) didn't always mean losing the war. It became clear that nothing galvanises public, political, media and financial support like a legal case, and that it often provides a central plank for multi-faceted and sophisticated campaigns. I've seen that play out many times over the last thirty years, be it in the field of air pollution, climate change or the licensing of the killing of wild birds.

I've also seen, over that period, ongoing reforms to the process of JR that first (with the advent of the Aarhus costs regime) turned liberalising access to the courts from a pipedream to a realistic prospect. Since 2015, Governments in power have sought to incrementally emasculate such access – death by a thousand cuts. The time to embed the full extent of environmental rights in the Aarhus Convention (the right to know, the right to engage and the right to challenge) in a coherent piece of primary legislation is long overdue. Without that, the achievement of the central objective of the Convention – the ability of present and future generations to enjoy access to a healthy environment – will be hampered. If civil society is to continue playing a pivotal role in meeting the climate, biodiversity and energy crises we face, individuals, community groups and eNGOs must be properly equipped to do so.

Notes

1 The author is contracted by the RSPB to work on access to environmental justice and co-Chairs the Legal Strategy Group of Wildlife and Countryside Link in that capacity. Her remaining time is spent as a Consultant Solicitor for Leigh Day.
2 Daniel B. Hayhow, Mark A. Eaton, Andrew J. Stanbury et al, *The State of Nature 2019* (The State of Nature partnership 2019).
3 Daniel B. Hayhow, Fiona Burns, Mark A. Eaton et al, *State of Nature 2016* (The State of Nature partnership 2016).
4 Will Steffen, Katherine Richardson, Johan Rockström et al, 'Planetary Boundaries: Guiding Human Development on a Changing Planet' (2015) 347(6223) *Science* 736.
5 RSPB and Friends of the Earth, *A Pillar of Justice: The Impact of Legislative Reform on Access to Justice in England and Wales under the Aarhus Convention* (2019). https://policy.friendsoftheearth.uk/sites/default/files/documents/2020-01/A%20Pillar%20of%20Justice_.pdf
6 The Rt Hon Lord Bingham of Cornhill KG, 'The Rule of Law', lecture given on 16 November 2006 at the Centre for Public Law, University of Cambridge, https://www.cpl.law.cam.ac.uk/sir-david-williams-lectures/rt-hon-lord-bingham-cornhill-kg-rule-law.
7 *R (Alconbury Developments Ltd) v Secretary of State for the Environment, Transport and the Regions* [2001] UKHL 23.
8 UNECE Convention on Access to Information, Public Participation in Decision-Making and Access to Justice in Environmental Matters of 25 June 1999 (the Aarhus Convention), United Nations, Treaty Series , vol. 2161, p. 447.
9 Ibid., Article 9(4).
10 Ibid., Article 9(2).
11 *Wheeler v. Leicester City Council* [1985] AC 1054, 1078 B-C.
12 In contrast to the situation in Ireland, for example, where a general third party right of appeal exists.

13 *Council of Civil Service Unions v Minister for the Civil Service* A.C. 374, 410 (Diplock L)
14 Lady Justice Arden, 'Proportionality: The Way Ahead?' (2013) 3 *Public Law* 498–518.
15 *R (Andrew Cawdron) v North Norfolk Council and Balfour Beatty Civils Ltd* (unreported).
16 Council Directive 92/43/EEC of 21 May 1992 on the conservation of natural habitats and of wild fauna and flora, OJ L 206.
17 *R (on the application of (1) Derek Foster (2) Tom Langton (claimants)) v Forest of Dean District Council (Defendant) & (1) Homes & Communities Agency (2) Natural England (Interested Parties)* [2015] EWHC 2648 (Admin).
18 Case C-258/11 *Sweetman v. An Bord Pleanála*, EU:C:2013:220.
19 *Smyth v Secretary of State for Communities and Local Government* [2015] EWCA Civ 174, [79]–[80].
20 *Abbotskerswell Parish Council v (1) Teignbridge District Council (2) Secretary of State for Communities & Local Government* [2014] EWHC 4166 (Admin).
21 *R (on the application of Richard McMorn) v Natural England & Defra* [2015] EWHC 3297 (Admin).
22 *R (on the application of Dillner) v. Sheffield City Council* [2016] EWHC 945 (Admin).
23 *R (on the application of Mynydd v Gwynt Ltd) v Secretary of State for Business, Energy and Industrial Strategy* [2018] EWCA Civ 231.
24 *R (on the application of RSPB) v Natural England and R (on the application of Mark Avery) v Natural England & (1) Jemima Parry-Jones (2) Secretary of State for the Environment, Food & Rural Affairs* [2019] EWHC 585 (Admin).
25 At which time the EPLS ceased.
26 Francis Jacobs, 'The Role of the European Court of Justice in the Protection of the Environment' (2006) 18(2) *Journal of Environmental Law* 185–205.
27 *R (on the application of Lumsdon and others) v Legal Services Board* [2015] UKSC 41 [37]-[73].
28 See Findings and recommendations with regard to communication ACCC/C/2008/33 concerning compliance by the United Kingdom of Great Britain and Northern Ireland, adopted by the Compliance Committee on 24 September 2010.
29 Communication to the Aarhus Convention Compliance Committee concerning compliance by the United Kingdom of Great Britain and Northern Ireland with the provisions of the Convention on access to justice concerning the review of substantive legality (ACCC/C/2017/156).
30 *R (Spurrier & Others) v Secretary of State for Transport* [2019] EW HC 1070 (Admin)
31 *R (Mott) v Environment Agency* [2016] EWCA Civ 564; [2016] 1 WLR 4338 and *R (Mott) v Environment Agency* [2018] UKSC 10.
32 *R (Lumsdon) v Legal Services Board* [2015], supra n 27.
33 *Smyth v Secretary of State for Communities and Local Government* [2015], supra n 19.
34 *Plan B Earth v Secretary of State for Transport* [2020] EWCA Civ 214, at para. 75.
35 Ibid.
36 *Friends of the Earth Ltd & Ors, R (on the application of) v Heathrow Airport Ltd* [2020] UKSC 52.
37 *R (on the application of Christopher Packham) v (1) Secretary of State for Transport (2) The Prime Minister & High Speed Two (HS2) Limited (Interested Party)* [2020] EWCA Civ 1004.
38 Supra n 29.
39 Findings and recommendations with regard to communication ACCC/C/2013/90 concerning compliance by the United Kingdom of Great Britain and Northern Ireland. Adopted by the Compliance Committee on 26 July 2021.
40 Supra n 17.
41 James Mathew 1830–1908 Irish judge: R. E. Megarry Miscellany-at-Law (1955)
42 *(1) WWF-UK (2) RSPB (Petitioners) v (1) Scottish Natural Heritage (2) The Secretary of State for Scotland (3) The Highland Council (4) Cairngorms Chairlift Company (5) Highland & Islands Development Board* (Respondents) (1999) Env LR 632.

43 The complaint was drafted by the present author and submitted by the Coalition for Access to Justice for the Environment (CAJE), which at that time comprised WWF, the RSPB, Friends of the Earth, Greenpeace, the Environmental Law Foundation and Capacity Global.

44 Directive 2003/35/EC of the European Parliament and of the Council of 26 May 2003 providing for public participation in respect of the drawing up of certain plans and programmes relating to the environment and amending with regard to public participation and access to justice Council Directives 85/337/EEC and 96/61/EC, OJ L156.

45 Judgment of 13 February 2014 in Case C-530/11 *Commission v UK,* EU:C:2014:67.

46 Articles 143 and 144 of the Consolidated version of the Rules of Procedure of the Court of Justice of 25 September 2012 and Articles 139 and 140 of the Rules of Procedure of the General Court of 4 March 2015, OJ 2015 L 105, p. 1.

47 Communication ACCC/C/2008/33 United Kingdom.

48 *R (Corner House) v Department of Trade and Industry* [2005] EWCA Civ 192.

49 Findings and recommendations with regard to communication ACCC/C/2008/33 concerning compliance by the UK, supra n 28.

50 Ibid., 128.

51 Ibid., 132.

52 Ibid., 135.

53 Secondary Legislation Scrutiny Committee, 25th Report and a Motion of Regret regarding SI 2017/95 (L.1) laid by Lord Marks of Henley-on-Thames in the House of Lords on 14 March 2017 and debated on 13 September 2017, resulting in a defeat for the Government.

54 Figures taken from information obtained from the Ministry of Justice by the author under the Environmental Information Regulations 2004 and Ministry of Justice Civil Justice Statistics Quarterly, England and Wales, October to December 2018.

55 Civil Procedure Rules, Part 45.44.

56 Ministry of Justice, 'Costs Protection in Environmental Claims: Proposals to Revise the Costs Capping Scheme for Eligible Environmental Challenges' (2015) https://consult.justice.gov.uk/digital-communications/costs-protection-in-environmental-claims/supporting_documents/costprotectioninenvironmentalclaimsonsultationpaper.pdf.

57 Case C-530/11 *Commission v UK* EU:C:2014:67 and Case C-260/11 *The Queen, on the application of Edwards v Environment Agency, First Secretary of State, Secretary of State for Environment, Food and Rural Affairs,* EU:C:2013:221.

58 Findings and recommendations with regard to communication ACCC/C/2013/85 and ACCC/C/2013/86 concerning compliance by the United Kingdom of Great Britain and Northern Ireland. Adopted by the Compliance Committee on 17 June 2015.

59 Report of the Compliance Committee on compliance by the United Kingdom of Great Britain and Northern Ireland – Part I to the Seventh Meeting of the Parties to the Convention, para. 30. https://unece.org/sites/default/files/2021–09/ece.mp_.pp_.2021.59_ac.pdf

60 *(Bertoncini) v London Borough of Hammersmith and Fulham and Kendall Massey* CO/3213/2019 [2020] EWHC

61 Civil Procedure Rule 45.43(4).

62 *R v Swale Borough Council, ex parte Royal Society for the Protection of Birds* (1990) 2 Admin LR 790.

63 Judgment of 17 October 2013 in Case C-533/11 *European Commission v Belgium,* EU:C:2013:659.

64 Ministry of Justice Civil Justice Statistics Quarterly, England and Wales, October to December 2021.

65 *R (on the Application of Thomas Langton v (1) The Secretary of State for Environment, Food and Rural Affairs and (2) Natural England* [2019] EWHC 597 (Admin).

66 Ibid., 44.

67 As of 1 July 2013, the Civil Procedure (Amendment No. 4) Rules 2013 reduced the time limit for filing a claim from three months to six weeks for decisions made under the 'Planning Acts' (i.e. the Town and Country Planning Act 1990, the Planning (Listed Buildings and Conservation Areas) Act 1990, the Planning (Hazardous Substances) Act 1990 and the Planning (Consequential Provisions) Act 1990.
68 Case C-669/16 *Commission v UK,* EU:C:2018:844.
69 Part Two (Trade [etc] Arrangements) Heading One (Trade), Title XI (Level playing field), Chapter 7 (Environment and Climate).
70 Section 8 Environment Act 2021.
71 Section 17 Environment Act 2021.
72 Dr Ronan Cormacain, *Environment Bill – Office for Environmental Protection: A Rule of Law Analysis.* (The Bingham Centre for the Rule of Law 2021). https://binghamcentre.biicl.org/publications/environment-bill-office-for-environmental-protection-a-rule-of-law-analysis
73 Section 38(3)(6) Environment Act 2021.
74 Section 38(9) Environment Act 2021.
75 Supra n 72.
76 Office for Environmental Protection, 'The OEP Strategy and Enforcement Policy', consultation opened 25 January 2022 and closed 22 March 2022, https://consult.theoep.org.uk/oep/the-oep-strategy-and-enforcement-policy/.
77 Department for Environment, Food and Rural Affairs, 'Environment Act 2021: Environmental targets', published 16 March 2022, https://www.gov.uk/government/consultations/environment-act-2021-environmental-targets.
78 'A Green Future: Our 25 Year Plan to Improve the Environment', sets out what we will do to improve the environment, within a generation. https://www.gov.uk/government/publications/25-year-environment-plan
79 Wildlife and Countryside Link, 'Government Must Aim Higher on Environment Targets', 16 March 2022, https://www.wcl.org.uk/government-must-aim-higher-on-environment-targets.asp.
80 Defra, *Biodiversity 2020: A Strategy for England's Wildlife and Ecosystem Services. Indicators* (2021). https://assets.publishing.service.gov.uk/government/uploads/system/uploads/attachment_data/file/1058728/England_biodiversity_indicators_2021_FINAL_REVISED_version_3.pdf.
81 In January 2022, the Prime Minister announced that there would be a new 'Brexit Freedoms' Bill to end the special status of EU law and ensure that it can be more easily amended or removed. This will form part of a major cross-government drive to cut £1 billion of red tape for businesses and improve regulation. https://www.gov.uk/government/news/prime-minister-pledges-brexit-freedoms-bill-to-cut-eu-red-tape.
82 Professor Maria Lee, Faculty of Laws, University College London.
83 Sections 112 and 113 Environment Act 2021.
84 Department for Environment, Food and Rural Affairs, 'Nature Recovery Green Paper: Protected Sites and Species', consultation opened 16 March 2022 and closed 11 May 2022, https://consult.defra.gov.uk/nature-recovery-green-paper/nature-recovery-green-paper/.
85 Don C. Smith, 'Environmental Courts and Tribunals: Changing Environmental and Natural Resources Law Around the Globe' (2018) 36(2) *Journal of Energy & Natural Resources Law* 137–140.
86 George (Rock) Pring and Catherine (Kitty) Pring, 'Environmental Courts & Tribunals: A Guide for Policy Makers' (2016) *UN Environment Programme.*
87 Ibid.
88 Brian J. Preston, 'Characteristics of Successful Environmental Courts and Tribunals' (2014) 26 *Journal of Environmental Law,* 365, 387.
89 Presentation given by the Hon. Brian J Preston SC, Chief Judge of the Land and Environment Court of New South Wales, Australia at an event hosted by the UK

Environmental Law Association on 'An Environmental Court after BREXIT' on 24 July 2018 and subsequent personal communication 30 July 2018.

90 Ibid.

91 See HM Courts & Tribunal Service, 'Environmental Fines or Notices: Appeal Against a Regulator' (17 November 2014) https://www.gov.uk/guidance/environmental-fines-or-notices-appeal-against-a-regulator.

92 See, for example: (i) Malcolm Grant, 'The Environmental Court Project – Final Report to the Department of Environment, Transport and the Regions' (1999); (ii) Michael Woods and Richard Macrory 'Environmental Civil Penalties: a More Proportionate Response to Regulatory Breach' (2003) Centre for Law and the Environment, Faculty of Laws, University College London; and (iii) The Environmental Justice Project (2004) 'Environmental Justice', https://unece.org/DAM/env/pp/compliance/C2008–23/Amicus%20brief/AnnexCEJP.pdf.

93 Environmental Rights Centre for Scotland. 2021. Why Scotland needs an environmental court or tribunal. https://www.ercs.scot/wp/wp-content/uploads/2021/12/Why-Scotland-needs-an-ECT-Oct-2021.pdf

94 For example, the Swedish Land & Environment Courts, the Land and Environment Court of New South Wales, and the New Zealand Environment Court. For a thorough examination of the issue, albeit now somewhat dated, please see Malcom Grant's Final Report on Environmental Courts (1999), ibid.

95 The commencement of proceedings in the Land and Environment Court of New South Wales may or may not have a "suspensory effect" depending on the nature of the proceedings. Further information about injunctions in environmental cases, including interlocutory injunctions is provided in: Brian J. Preston, 'Injunctions in planning and environmental cases' (2012) 36 *Australian Bar Review* 84–152.

96 Both of these remedies are exercised by the New Zealand Environment Court – see Malcolm Grant (1999), supra n 92.

97 In the South Australia Environment Court, the power to award exemplary damages may be exercised only by a judge. Damages are assessed having regard to any detriment to the public interest resulting from the breach, any financial or other benefit that the respondent sought to gain by committing the breach, and any other matter the Court thinks relevant – see section 214 of the Planning, Development and Infrastructure Act 2016.

9

SCIENCE AND THE INTERPRETATION OF THE EU HABITATS DIRECTIVE BY THE COURT OF JUSTICE OF THE EU

2+2=5?

Hendrik Schoukens

Introduction: George Orwell, Thomas Kuhn and the Habitats Directive

George Orwell's infamous Ministry of Truth – which lies at the heart of his renowned novel 1984[1] – appears to be relatively off-chart when writing a chapter on the application of the Habitats Directive[2] in courtrooms. As of today, this directive is often perceived as one of the most effective pieces of international nature conservation legislation. More generally, environmental law tends to be portrayed as an area of law mainly characterised by objective facts and technical figures, with little room for emotions and subjectivism. This ties in with the idea of 'science' as an area characterised by objectivity and rationality. However, this perception is no longer in line with the existing practice witnessed in courtrooms. Whereas environmental cases are often presented as purely technical or factual, they increasingly give rise to heavily charged emotional debates on the adequacy of the evidence presented. Multiple truths emerge. Measurement and quantification are not capable of fully liberating scientific research from personal biases. It was Thomas Kuhn, the famous American philosopher of science, who claimed science to be subjective by nature.[3] Kuhn argued that paradigms and paradigm shifts were the central focal points when discussing the matter of the objectivity of science. Seeing that there are so many competing paradigms and differences of opinions, science is inescapably subjective. Since all scientific theories are based on subjective conditioning and ideas, there is no way that science can ever be fully objective.

This is precisely where Orwell's nightmarish vision of the future jumps in. As is generally known, the name *Ministry of Truth* was but a misnomer. It is a poignant illustration of what Orwell tags as 'doublespeak' because it serves the opposite: the falsification of historical events. Even more so, one of the Ministry's

DOI: 10.4324/9780429299100-9

principal tasks was the spreading of so-called 'Newspeak'. This is illustrated by the infamous equation: 2+2=5, which is supposed to be 'true' whenever the situation requires so. In an age where 'fake news' is increasingly used as a political tool to win elections or to question, during a global pandemic, the effectiveness of vaccination campaigns, it would be relatively easy to imagine that Orwell's dystopian vision of the future has now fully materialised. While deliberate disinformation and the dissemination of 'hoaxes' are certainly nothing new and have even spiked since the arrival of Covid-19, the rise of the internet and social media has created a perfect storm in which such discourses can thrive more than ever. When even high-profile politicians are setting aside criticism of their environmental policy as 'fake news' or openly questioning the scientific credentials of climate scientists, it should not come as a surprise to see a tightening of the debate in the courtroom as well. This has rendered the task of judges more difficult and complex when adjudicating environmental cases, which often present highly technical, factual and scientific questions. Equally in the context of EU nature conservation law. For instance, how is one to assess the favourable conservation status for a species such as the Grey wolf (*Canis lupus*), which was exterminated in many countries several centuries ago? How to assess the resilience of ecosystems that will be affected by project developments in times of global heating?

Polish old-growth forests, a Belgian transport company and the Swedish wolf hunt: science vs the rule of law?

To further illustrate the main thesis of this chapter, I first delve into some concrete legal proceedings where science was used (or abused?) when implementing EU nature conservation law.

As of today, the relatively recent court battle regarding the protection of the Polish Białowieża forest, one of Europe's most pristine primeval forests, probably represents one of the most compelling illustrations of this recent tendency to frame scientific studies into legal arguments pro or contra potentially harmful activities. Some years ago, State Forests, the Polish organisation competent for managing most Polish forests, put forward its highly controversial logging policy. It claimed that Białowieża forest was threatened by an outbreak of the spruce bark beetle, an insect that kills weak spruce trees. Massive logging was presented as the logical solution for this threat. Yet these plans gave rise to a sharp conflict that has divided Poland and pitted foresters against ecologists and other researchers. In their protests, environmental organisations and groups reacted to what they viewed as an unprecedented slashing of one of the EU's most valuable forests. Yet, when the CJEU was asked to issue a temporary halt to the logging, the Polish government justified its logging practices with reference to a so-called massive 'scientific' experiment, which had been carried out during the years preceding the court proceedings. In other words, 'science' was used to justify the Polish logging activities. Tellingly, this approach was strongly critiqued by other scientists as mere 'propaganda' by the Polish government, with one Polish ecologist openly comparing State Forests to 'Orwell's Ministry of Truth'.[4] Such scientists claimed 'science' backed their viewpoints, by underlining the ecological importance of dead wood and untouched natural habitats of many endangered species.

It would be easy to discount or downplay this Polish case as an exceptional event. It is not. Numerous other examples of scientific controversy impacting the outcome of environmental decision-making procedures have emerged elsewhere in the European Union. For instance, when the Flemish logistics service supplier H. Essers intended to expand its premises in Genk into an adjacent Natura 2000 site in 2015, ecological considerations were only of ancillary importance during the decision-making process. This led to one of the most mediatised court cases in the Flemish region regarding the application of the Habitats Directive. Several environmental NGOs claimed that the expansion zone, which was located inside a protected Natura 2000 site, harboured some very valuable natural habitats and species.[5] Interestingly, the Flemish government decided to consult several scientists with but one clear goal: gathering 'objective' evidence to prove that the forest to be destroyed contained little valuable biodiversity. And thus its disappearance would not be of major relevance for the protected area. One university professor was asked to draft a report in which he downplayed the ecological relevance of the site.[6] This report concluded that the area contained some interesting ecological potentialities. However, its disappearance would not harm the overall integrity of the site. The report was subsequently leaked to the press and presented as the ultimate argument to discredit the environmental protests against the logging. When confronted with this report, the environmental NGOs merely replicated this tactic. They in turn commissioned an ecological study at another university, which apparently came to an entirely different conclusion. The report stated that the forest contained valuable patches of habitat instrumental to achieving the conservation objectives of the site. This was used to illustrate the significance of the planned actions, leading the drafters to call for a ban on the proposed deforestation. The apparently conflicting studies were used as an additional argument to underline that significant doubt arose as to the absence of significant effects for the Natura 2000 site in question. Ironically, the two reports did not differ that much in terms of substance, since both underlined the restoration potential of the site. However, the way the reports were used is clearly representative of a more confrontational approach to science nowadays.[7]

A similarly heated scientific controversy arose in Sweden concerning the justification of the annual wolf hunt. As of today, this country currently hosts a population of about 400 wolves in central Sweden. Their number has increased substantially during the last decade yet the population is heavily inbred: all the wolves descend from a handful of animals that arrived from Finland since the 1980s. However, hunters and livestock farmers persistently urged the public authorities to curb the size of the growing population. Whereas a limited wolf cull to prevent serious damage to livestock appears to be permissible under Article 16(1)(b) of the Habitats Directive, the Swedish authorities were under political pressure to reduce the wolf numbers even more in order to appease certain lobby groups.[8] Again, 'science' was used as an instrument for a more flexible management approach. The concept of 'tolerance hunting' was put forward, which aimed to bolster support for the presence of large carnivores in the countryside by granting yearly permits to hunters allowing them to eliminate a selected number of wolves every year.[9] It was argued that scientific reports bolstered this thesis: culling wolves would, it was claimed, lead to greater societal acceptance. The idea is sometimes coined as 'killing wolves to save them'. More generally, the Swedish authorities claimed that in spite of the Grey wolves' genetic problem linked to inbreeding, the wolf population in its country had achieved a favourable conservation status (FCS). In doing so, they based their reasoning partially on a scientific report drafted by Guillaume Chapron, who focused on the population viability of the Swedish wolf population. To avoid misinterpretation of his work, which excluded genetic aspects, Chapron proactively underlined that the report was not to be used to estimate FCS, precisely because it excluded a genetic analysis.[10] Nevertheless, the Swedish government used this and other scientific reports in support of its claim that the wolf population had achieved FCS.[11]

Of course, the case studies presented above are not exhaustive. Still, they aptly capture the myriad of difficult questions at stake when dealing with natural sciences in the courtroom. In this chapter, I will further explore the approach used by the Court of Justice of the EU (CJEU) to interact with science when dealing with the application of the Habitats Directive, which, together with the EU Birds Directive,[12] constitutes one of the main pillars of the EU's environmental policy. In so doing, the chapter aims to unravel the complex plethora of issues at stake in the interface between science and law. After having presented some general reflections on the interface between science and law, I will posit that science and research are deeply ingrained in the Habitats Directive, and will be increasingly at the forefront in court cases as a result of the increased relevance of the 1998 Aarhus Convention[13] in EU law. Based on an analysis of recent jurisprudence of the CJEU, I argue that the increased reliance on the precautionary principle when adjudicating nature conservation cases, among other things, has allowed the EU court to stay at a safe distance from substantive scientific disputes concerning the ecological effectiveness of certain policy approaches. This pragmatic approach has given rise to some remarkable environmental wins on the ground. Nevertheless, I will also highlight that the CJEU's approach has certain limits and downsides.

Science, courts and the rule of law: an unlikely combination?

It would be relatively easy to discount the above-mentioned national cases regarding the inadequate use of science in courtrooms as anecdotal evidence. Against the backdrop of recent developments, any such discounting no longer appears to be justified. A few years ago, the French ecologist Guillaume Chapron, whose research was placed at the centre of the aforementioned heated debate concerning the acceptability of the Swedish wolf hunt in light of the Habitats Directive, openly cautioned against the tendency to abuse science when implementing nature protection laws, in a 2014 contribution in *Nature*. In unusually sharp language, Chapron laid bare the increasing challenges scientists face when they are caught up in a political debate on the acceptability of certain activities, such as hunting and intensive agriculture:

> I am concerned that Sweden's misuse of my research and its flouting of European regulations will set a dangerous precedent in biodiversity conservation. The distortion of science has been very subtle and technical in this case, and the wolves will not be eradicated, but it is important to highlight because it may be the first of many examples.[14]

Chapron's words resonate. The potential misuse of science is not making the task of judges any easier when it comes to adjudicating environmental cases. In the literature, it is highlighted that such adjudication faces two major challenges: an absence of scientific expertise among judges, and vague norms.[15] Courts, which have a natural tendency to look for manifest cases of non-compliance, oftentimes appear ill-equipped to deal substantively with environmental cases because of an apparent lack of technical expertise. For now, judges are trained to assess legal claims in view of the 'rule of law', which relies on criteria such as clarity, predictability and consistency.

In a recent contribution, Tracey Kanhanga further elaborates on the multiple challenges that lie ahead when it comes to the role of science in environmental disputes.[16] In her analysis, Kanhanga quotes Weinberg, who brilliantly captured the difficult dilemma many judges are currently facing when dealing with environmental cases. In particular, Weinberg posited that the judge's dilemma is deeply rooted in the underlying ideas of our legal system. He claims that the law 'expects a judge to decide cases when science is uncertain, when science can't help him'. Or, put differently, a judge is required to exert his judicial powers 'outside the limits of legal certainty'.[17] Likewise, Mangel asserts that judicial decisions are predicated on the translation of scientific information into a form appropriate to render well-reasoned and fair judicial decisions.[18] But when science is inconclusive, it can become burdensome to come forward with clear-cut regulations or decisions. Even more so, the question remains what is to be regarded as a 'fair outcome'. What if the lack of scientific certainty is used by a court to call into question the legality of authorisations for a country's most vital economic assets, such as intensive agriculture or harbour activities? What if science is in flux, which might give rise to different legal duties? Is a simple reference to the precautionary principle sufficient to refuse permit applications when conflicting scientific studies arise? And how is a judge to assess what level of 'ontological' doubt would still be permissible?

The deeply ingrained desire to expect clear-cut legal answers in a context where science is failing to produce definitive answers might give rise to judicial self-restraint, as evidenced by the jurisprudence of the International Court of

Justice (ICJ).[19] In the 1997 *Gabčíkovo-Nagymaros* case, for instance, in which the ICJ had to deal with conflicting scientific reports produced by Hungary and Slovakia, the Court notoriously opted to not engage with the scientific and technical evidence presented to it by the states, because assessing the adequacy of the reports was not necessary to answer the legal questions put before it.[20] This pragmatic attitude is not unique and is indeed often repeated at the national court level. Rather than ending up entangled in scientific debate, courts have a natural tendency to focus on the legal issues at stake. However, when the ICJ explicitly decided to weigh and assess the large body of scientific studies put before it in the 2011 *Pulp Mills* case, this bold move was vehemently opposed by several judges, who recorded their discontent in dissenting opinions. According to judges Al-Khasawneh and Simma:

> (t)he adjudication of disputes in which the assessment of scientific questions by experts is indispensable, as is the case here, requires an interweaving of legal process with knowledge and expertise that can only be drawn from experts properly trained to evaluate the increasingly complex nature of the facts put before the court.[21]

This quotation perfectly illustrates the myriad challenges in the context of environmental litigation, both at international and national court levels. Interestingly, when having to decide on the legality of Japan's whaling programme in the Antarctic, the ICJ had to shed light on its interpretation of the concept of 'science'. Japan claimed its permits were legal since they had been issued 'for the purpose of scientific research' within the meaning of Article VII of the 1946 Whaling Convention. However, the ICJ's approach appears ambivalent at best. For one, it openly contested that its role was 'to resolve matters of scientific or whaling policy' at all.[22] Likewise, it could not be compelled to 'pass judgment on the scientific merit or importance of [Japan's programme] nor (...) decide whether the design and implementation of a programme are the best possible means of achieving its stated objectives'.[23] Although the ICJ indirectly considered issues such as the role of a testable hypothesis in scientific research, the importance of peer review, and how models and data work, it did not put forward an alternative set of criteria to determine whether reports constitute 'science', underlining that 'the Court [does not] consider it necessary to devise alternative criteria or to offer a general definition of scientific research'.[24] Echoing the previous dissenting opinions in the *Pulp Mills* and *Gabčíkovo-Nagymaros* cases, Judge Owada (in another dissenting opinion) rephrased earlier criticism regarding the inaptness of the court to deal with scientific disputes by stating that the ICJ

> as a court of law, is not professionally qualified to give a scientifically meaningful answer, and should not try to pretend that it can, even though there may be certain elements in the concept that the court may legitimately and usefully offer as salient from the viewpoint of legal analysis.[25]

However, in spite of the prevalent ambivalence, I concur with Mangel that a substantive analysis of the latter judgment underscores that the ICJ judges clearly 'understood the links between objectives of a scientific study, methods to achieve those objectives, and the importance of appropriate sample size'.[26]

Mangel rightly underlines the importance of judges who lack formal scientific training building the necessary capacity to understand the scientific process rather than appointing experts to unravel the scientific riddles placed before them. Even so, the ICJ's constitutive instruments allow it to have recourse, when necessary, to experts to assess the scientific evidence placed before it. This was poignantly illustrated in the *Costa Rica v. Nicaragua* and the *Nicaragua v. Costa Rica* cases in 2016, where the ICJ eventually decided to call upon technical experts.[27] Yet even more progressive approaches appear possible, with Article 50 of the ICJ rules allowing the ICJ to entrust other parties more suited to fact finding to give an expert opinion on scientific matters that are adduced before it.[28] As a result, the Court could appoint competent scientists to review complex environmental reports presented before it by the parties. Still, the question remains whether the appointment of technical experts will enable judges to clarify long-standing scientific debates, especially in cases where both sides claim to have 'science' on their side.

Why science is increasingly important for the effective implementation of the Habitats Directive: substantive and procedural arguments in the mix

In contrast with the above analysis regarding the interface between science and law at international level, the CJEU has not yet been explicitly asked to pronounce on the basic fundamental questions regarding science and the rules thereof, which have emerged in the above-mentioned jurisprudence of the ICJ. At least, that is the widely shared impression when approaching the recent case law of the CJEU. However, this does not mean that the CJEU has not been confronted with similar challenges. Given its ecological ambitions, the Habitats Directive, which stands out as one of the most 'popular' EU environmental directives to be invoked before the CJEU, is presenting the CJEU with ever-increasing opportunities to elaborate on the role of science when implementing the conservation and restoration duties incumbent upon the Member States by virtue of the Habitats Directive.

Before delving into this interesting body of case law, it is worthwhile pinpointing the two main reasons for this recent surge in scientific debates in nature protection-based cases before the CJEU.

Science and conservation: two sides of the same coin?

Adopted in 1992, the Habitats Directive protects over 1,000 European species other than birds and about 200 natural habitats which are considered to be endangered, vulnerable, rare and/or endemic. Protected sites – which are part of an EU-wide ecological network dubbed 'Natura 2000' – are to be designated to ensure that the natural habitats and species listed in the Habitats Directive reach a favourable conservation status within the EU, which implies that in some instances robust recovery efforts will be required. These designation duties are listed in Article 4 of the Habitats Directive, while Article 6 contains the basic conservation duties in respect of these protected sites.[29] For several endangered species a strict protection regime is set out in Articles 12–16 of the Habitats Directive.[30]

The Habitats Directive itself, however, makes explicit the importance of science for the implementation of its conservation and recovery duties. This is most notable in Article 18, which states that Member States and the Commission 'shall encourage the necessary research and scientific work having regard to the objectives set out in Article 2 of the Habitats Directive and the monitoring obligation included in Article 11'. Paragraph 2 of the same article underlines that particular attention is to be paid to scientific work necessary for the implementation of the Natura 2000 network (Article 4) as well as for the establishment of ecological corridors (Article 10). Also, transboundary cooperative scientific research is to be encouraged.

Beyond this explicit provision, science also comes into the picture for other core duties of the Habitats Directive. The seminal concept of 'favourable conservation status' (FCS), which is defined in Article 1(e) and 1(i) of the Habitats Directive, is heavily dependent on scientific input. To give but one example: in order to determine whether a species has reached a favourable conservation status (FCS), the population dynamics and the evolution of the natural range of the species need to be examined. This clearly requires scientific research by ecologists on population dynamics and conservation threats. The same can be said when it comes to the provisions on site protection.[31]

In addition to the references to science included in Article 18, Article 4(1) of the Habitats Directive explicitly requires Member States to take into account 'relevant scientific information' when proposing a list of sites to be included in the Natura 2000 network. Furthermore, the duty to present six yearly implementation reports in Article 17 of the Habitats Directive mandates the engagement of the scientific community since it entails an overview of the measures taken pursuant to Article 6(1) of the Habitats Directive as well as their impact on the conservation status of the natural habitats and species in question. Lastly, scientific research also plays a crucial role when applying the rigid procedures for amending the listed habitats and species on Annexes I, II, and V (qualified majority) and IV (unanimity).

Science before courts: the bumpy road towards effective judicial protection

This brings me to a second preliminary observation, which concerns recent developments in the field of access to justice in environmental cases. In my view, the incremental references to science in cases concerning the application of the Habitats Directive cannot be dissociated from the rising relevance of the third pillar of the Aarhus Convention, which has been ratified both by the EU as well as the Member States and thus constitutes an integral part of the acquis. To be more precise, Article 9(2) and 9(3) of the Aarhus Convention in particular oblige national courts to provide members of the public, including environmental NGOs, with relatively broad access to justice in environmental cases. These provisions, which have been partly implemented by EU secondary legislation,[32] have particular relevance for the application of the Habitats Directive. This is especially the case since the Habitats Directive, in contrast to the Environmental Impact Assessment (EIA) Directive, does not contain an explicit provision regarding access to justice.

Two remarks are in order here.

First, there is the increased importance of effective judicial protection in the context of private enforcement of environmental duties before national courts. Recent case law developments of the CJEU have clearly underlined the relevance of the Aarhus rationale in the context of legal proceedings which are based on the conservation duties laid down in the Habitats Directive. For instance, in its landmark ruling in the *Slovakian Brown Bear* case in 2011, the CJEU's refusal to accord direct effect to Article 9(3) of the Aarhus Convention did not prevent it from underlining the importance of effective environmental protection at national level, even in the context of court cases regarding EU protected species.[33] From that moment onwards, the train of progressive case law regarding access to justice in environmental cases appears unstoppable. In the *Trianel* case, which related to a situation covered by access to justice provisions included in the EIA Directive, the CJEU clearly underlined that environmental NGOs by definition have rights and interests to protect the environment. When these 'rights' are infringed, they must have wide access to justice to seek judicial review.[34] This line of jurisprudence has been reasserted in more recent case law.[35] What is more, in a 2016 follow-up to its 2011 decision in the *Slovakian Brown Bear* case, the CJEU also determined that Article 9(2) of the Aarhus Convention is also relevant– via the detour of Article 6(1)(b) of the Aarhus Convention, and the right to participate in decisions subject to Article 6(3) of the Habitats Directive – when authorising projects with a potential adverse effect on protected Natura 2000 sites. Importantly, the CJEU noted that the lack of a specific provision on access to justice in the Habitats Directive was not decisive when the right to an effective remedy, as stated in Article 47 of the Charter of Fundamental Rights, already requires national courts to grant effective access to justice to environmental NGOs in nature conservation cases.[36] Wide *locus standi* is thus provided to environmental NGOs through both EU law and the Aarhus Convention, which provides increased opportunities for environmental litigation, with the possibility of scientific controversies being aired.

Second, the recent surge of disputes relating to science in nature conservation cases is equally linked to the increased importance of substantive legality review in the context of EU environmental law. Evidently, the intensity of judicial review might vary from one jurisdiction to another depending on the nature of the procedure (be it an administrative or a civil proceeding). However, there appears to be a clear shift from a formal legality review towards a more in-depth review of decisions and/or omissions in light of the substantive arguments raised by individuals or environmental NGOs.[37] For both Article 9(2) and Article 9(3) of the Aarhus Convention do not merely oblige national courts to grant wide access to justice in environmental cases to environmental NGOs but also require Member States to ensure an effective judicial review of both the procedural and the substantive legality of decisions, acts and omissions falling within the scope of these provisions.[38] Even though Article 9(3) of the Aarhus Convention does not contain an explicit reference to substantive review, the case law of the CJEU has nevertheless put forward the principle of effective judicial review covering substantive and procedural legality of decisions and omissions falling under Article 9(3) of the Aarhus Convention.[39] As a result, criticism regarding the adequacy and the content of EIAs for project developments prone to adversely affect Natura 2000 sites falls squarely within the scope of judicial review before national courts. This was further underlined by the CJEU in its 2012 ruling in the *Altrip* case, where it concluded that excluding the applicability of judicial review in cases in which an EIA is found to be vitiated by defects would render the provisions regarding public participation in the EIA Directive largely ineffective.[40] In other words, national courts are required to critically review scientific reports put before them that are capable of invalidating the outcome of EIAs. Simply holding that a further review of the scientific underpinning of planning permits falls beyond the scope of the court's competence therefore no longer suffices.

Importantly, such an approach is not confined to secondary EU legislation, such as the EIA Directive, which explicitly requires a review of substantive and procedural legality, but has instead been promoted to a general requirement of EU environmental law. Pursuant to the 2017 Communication of the European Commission on Access to Justice, a 'national judge may (...) be called upon to take into account the relevant scientific evidence on which environmental measures are normally based'.[41] This is specifically relevant in the context of nature conservation cases. It draws on the CJEU's ruling in the 2004 *Waddenzee* case, which is addressed in more detail below.[42] The European Commission rightly concluded that the *Waddenzee* test cannot be treated by the national court as 'a subjective one lying exclusively within the public authority's own discretion', which further opens the door for environmental NGOs to bring forward scientific evidence pointing out flaws or loopholes present within the scientific and environmental reports on which permitting decisions are based.[43]

The duties of national courts in this regard were recently clarified by the Aarhus Convention Compliance Committee,[44] which held that,

> 119. [...] article 9(2) [of the Aarhus Convention] requires that a reviewing body must review all the facts, evidence and arguments before it and, based on that review, determine whether the contested decision is lawful. This requires the court to carry out its own assessment, in the light of all the

> evidence before it, as to whether the applicable legal requirements were met. The court must also clearly set out its reasoning when doing so.
>
> 120. For example, in a challenge to the substantive legality of an EIA screening decision, the court must make its own assessment, based on all the evidence put before it, as to whether the proposed activity was likely to have a significant effect on the environment and thus to require an EIA. It would not be sufficient to merely check that the decision-maker carried out the correct procedural steps for determining whether a project was likely to have significant effects. Nor does it suffice for the court to check that the decision-maker had formally applied the correct legal test and that the decision-maker had convinced itself that that test was met in a particular case.
>
> 121. To be clear, the Convention does not require the court to undertake a completely fresh analysis of all matters arising in the case and to substitute its decision for the decision taken by the competent authority. Nevertheless, the court must undertake its own assessment of all the evidence before it to determine whether the applicable legal requirements were met. The Committee considers that this requires the court to perform a review function over findings of fact and the weight to be given to evidence where those may have a direct impact on the determination as to whether the applicable legal test (for example, likely significant effects) has been met.

As revealed by the CJEU's jurisprudence in the 2008 *Janecek*[45] and the 2015 *ClientEarth* cases,[46] the same rationale may also apply outside the context of project-review and subsequent legal proceedings. The cited case law was linked to the requirement to prepare air quality plans under the EU Air Quality Directive 2008/50/EC. In this context, the CJEU held that environmental NGOs and individuals directly concerned can sue national authorities and ask courts to scrutinise the adequacy of the measures included in such plans, an approach which is reiterated in the context of the National Emission Ceilings Directive 2016/2284/EU.[47] The main thread of the mentioned case law therefore appears to be that national courts are competent to review the legality of national plans aimed at achieving objectives of EU environmental law. Equally, they are obliged to order authorities to adopt the required protection measures in light of the overarching principle of effective judicial protection if needed in view of the applicable substantive norms.[48] And thus, indirectly, scientific debates are increasingly imported by necessity into the courtroom as they are instrumental to reviewing the substantive underpinnings of actions implementing the Habitats Directive.

Science and the CJEU: lessons from recent case law regarding the Habitats Directive?

Given that the traditional judicial deference in handling scientific evidence is no longer allowed when the conservation of EU-protected nature is at stake, the question arises as to how national courts should proceed. In this section,

I focus on the main guidelines provided by the CJEU when interpreting the Habitats Directive in the context of scientific controversy. Most of the cases addressed below arose from preliminary references from national courts seeking to review whether public authorities had correctly applied the Habitats Directive. However, some of the cases are drawn from infringement proceedings, where the European Commission sued Member States for their non-compliance with the Habitats Directive.

An effective ecological network: ecological science at the forefront

An obligatory first step when assessing the handling by the CJEU of science in the context of the Habitats Directive is its case law with respect to the designation duties regarding the Natura 2000 network.[49] In one of the first landmark rulings in this regard, concerning the Santoña marshes in Spain, the CJEU in 1993 held that, while Member States retain a certain margin of discretion as regards the choice of Special Protection Areas (SPAs), the classification of these sites is directly subject to ecological criteria.[50] In other words, when a Member State is faced with cogent scientific evidence that a site is of particular interest for the conservation of certain endangered bird species, it would be failing to comply with its duties under Article 4(1) of the EU Birds Directive if it declined to select this site as part of the Natura 2000 network. Member States are required to designate sites that meet the ecological criteria, even when such designation is prone to cause economic impact.[51] In other words, scientific evidence should trump economic arguments when deliberating the selection of certain vulnerable areas. In retrospect, this represented the first clear-cut victory for science in the context of EU nature conservation law.[52] Its importance in terms of underlining science-based decision-making when dealing with endangered habitats cannot be overstated.

More recent case law is illustrative of a relatively open yet also remarkably pragmatic approach on the part of the CJEU toward scientific reports when assessing a Member State's compliance with its obligations. For instance, in its 1996 ruling in the *Commission v The Netherlands* case, the CJEU highlighted the importance of the *Inventory of Important Bird Areas in the European Community* (*IBA*), first published in the year 1989 by an environmental NGO in cooperation with Commission experts.[53] This exemplifies the Court's willingness to use scientific documents produced by environmental NGOs as an important benchmark in assessing whether Member States have effectively complied with their designation duties under the EU Birds Directive. The CJEU has repeatedly ruled that, although the IBA is not legally binding on the Member States, it can, by reason of its scientific value in these matters, be used by the Court as a basis of reference in this respect.[54] The CJEU did not find it necessary in any of these cases to rely on scientific experts to corroborate the scientific value of the IBA.[55]

That said, it would be incorrect to submit that the CJEU has treated the IBA as quasi-binding or determinative. In its other jurisprudence, the CJEU has clarified that the mere fact that a site is *not* included in the IBA is not in itself sufficient to justify non-designation, especially when other scientific reports demonstrate that the site in question complies with the ornithological criteria laid down in the Birds Directive. The CJEU has opted for the middle road. For instance, in its 2007 ruling on Ireland's implementation efforts under the EU Birds Directive, the CJEU further elaborated its view regarding use of the (updated) IBA. In this respect, the CJEU maintained that the mere fact that a Member State had not designated all the sites that were listed as important bird areas in the IBA was not to be regarded as conclusive in order to conclude that the former failed to comply with its obligations under the EU Birds Directive.[56] The CJEU equally indicated that due regard could also be given to other ornithological studies whenever they are objectively verifiable, in order to assess the adequacy of a Member States' designation efforts.[57] In other words, the IBA is to be used as a rebuttable presumption of compliance with Article 4(1) and (2) of the Birds Directive. The burden of proof seems to rest with the Member States, who have to demonstrate that a given non-designation is justifiable in view of other scientific reports.[58] It is impossible to justify the non-designation of a site which hosts important ecological values by reference to policy reports that lack scientific underpinnings.

So far, few Member States have succeeded in doing so. In many instances, Member States will grapple with providing sufficient counter-evidence in order to sway the CJEU. For instance, in the 2007 case mentioned above, Ireland failed to put forward any other ornithological criteria which were objectively verifiable, as compared with those used in *IBA 2000*, to serve as a basis for a different decision regarding the need for site designation.[59]

In other cases, the CJEU has exemplified its willingness to further engage in what Mauerhofer appropriately dubbed the 'battlefield of scientific facts'.[60] For instance, in the 2007 *Commission v Greece* case, the CJEU did not hesitate to condemn Greece over its failure to designate protected sites for specific bird species, even though the evidence put forward by the European Commission seemed to conflict with the report submitted by the Greek government.[61]

What makes the jurisprudence of the CJEU regarding the use of scientific reports even more compelling is the fact that the CJEU has underlined that the Member States are also required to re-evaluate their designation policies in view of continued monitoring of results and scientific studies.[62] In other words, the designation duty is not to be regarded as static but entails a flexible and continuing approach. And thus emerging scientific studies, which point to new opportunities for the classification of previously 'hidden' natural habitats that have not been included in the IBA, might be used as a legal yardstick to advocate additional designation of vulnerable sites. For example, in its prominent decision in the 2006 *Commission v Austria* case, the CJEU held that '(t)he obligation to classify is not limited by the state of scientific knowledge at any given time.'[63] To put it simply, whenever new reports reveal the presence of a significant number or surface area of EU-protected nature, new designation duties might arise.

It is apparent that most of the case law relates to protected sites that are to be designated by the Member States under the Birds Directive. However, *mutatis mutandis* the same rationale applies in the context of the Habitats Directive.[64] Two recent decisions stand out. They both revolved around the issue of Member States seeking to declassify sites that had been previously designated as Natura 2000 sites. In the notable 2014 *Cascina Tre Pini* case, the CJEU accepted the possibility of Member States declassifying selected sites, but only if, despite compliance with the conservation duties imposed by Article 6 of the Habitats Directive, the site in question had become irretrievably unsuitable to meet the objectives of the Habitats Directive, so that its designation no longer appears justified. Only when scientific reports demonstrate that the site no longer has a role to play within the wider Natura 2000 network does such removal appear legal.[65] This stringent approach was echoed by the outcome of the Dutch *Leenheerpolder* case. In these legal proceedings, the CJEU declared a decision of the Dutch government to reduce the surface area of a Natura 2000 site unlawful because it was not supported by scientific reports. The reduction was mainly founded on a political decision to reassess Dutch nature conservation policy. The CJEU underlined that a reduction in size of a designated Natura 2000 site is acceptable where 'on the basis of evidence of a scientific nature, an error that vitiated the relevant scientific information is made apparent'.[66] This case law would seem to pave the way for a more iterative designation procedure, which allows for the taking into account of new natural or human-made developments. In times of climate change, this might push national authorities to reconsider existing designations in the light of rising temperatures and changes in the distributions of species and habitats.

The concept of favourable conservation status: a science-based yardstick and path to recovery?

The increasing interplay between science and law has not remained limited to the context of the designation of the Natura 2000 sites. In recent years, the concept of 'favourable conservation status' (FCS) has presented the CJEU with another opportunity to express its view on scientific and other technical questions. This legal-ecological concept constitutes one of the building blocks of the Habitats Directive, the aim of which is to maintain or restore EU-endangered natural habitats and species to a favourable conservation status.[67] Pursuant to Article 1(i) of the Habitats Directive the conservation status of a species will be regarded as 'favourable' when

> population dynamics data on the species concerned indicate that it is maintaining itself on a long-term basis as a viable component of its natural habitat, the natural range of the species is neither being reduced nor is likely to be reduced for the foreseeable future and there is, and will probably continue to be, a sufficiently large habitat to maintain its populations on a long-term basis.

As hinted at in the previous sections, the application of the concept of FCS is predicated on a multitude of elements, which evidently require considerable scientific input in order to be identified and operationalised.[68] It is for ecological scientists to determine whether a species is 'a viable component of its natural habitat', what reference scenario is to be used to determine whether a population will be able to maintain itself 'on a long-term basis' and when a species is capable of 'maintaining itself'.[69] At least, that is the case when we assume that political and economic considerations are left out of the equation. However, this conclusion is sometimes countered by industry, politicians and lobbyists with reference to Article 2(3) of the Habitats Directive, which allows Member States to take into account economic and sociocultural considerations when implementing the directive.

As of today, however, the precise repercussions of the concept of FCS have become most apparent in the CJEU's case law regarding the strict protection duties in Article 12(1) of the Habitats Directive. These duties are mainly designed to protect endangered species, listed in Annex IV to the Habitats Directive, against direct impacts such as hunting, habitat encroachments and disturbance. In addition, they also require Member States to lay down concrete, coherent and coordinated species protection measures to protect these endangered species in the field.[70]

In its recent case law, the CJEU has increasingly used the concept of FCS as a strict benchmark to review a Member State's compliance with Articles 12 and 16 of the Habitats Directive. A first, albeit indirect, application of the notion of FCS was the Finnish Wolf case, in which the CJEU was asked to review the compatibility of the Finnish wolf hunt with Article 16 of the Habitats Directive. While doing so, the CJEU gave some implicit hints as to the concrete interpretation to be given to FCS when drafting hunting programmes. One of the questions that arose was to what extent hunting was still permissible when a species' population had not yet reached FCS. Although the CJEU underlined that the FCS of the populations of the species concerned in their natural range remains a necessary precondition in order to grant hunting derogations, it clarified that

> the grant of such derogations remains possible by way of exception where it is duly established that they are not such as to worsen the unfavourable conservation status of those populations or to prevent their restoration at a favourable conservation status.[71]

Reading between the lines, it is clear that such permits need to be science-based, as this appears to be the only objective manner to establish such lack of significant effects. However, what was even more striking was the dismissal of the Commission's view regarding the compliance of the Finnish practice with the requirement of maintaining a species at a FCS with reference to the scientific reports that had been presented to it.[72] These studies demonstrated that the number of Grey wolves present on Finnish territory had risen from between 110 and 130 specimens to between 185 and 200 specimens. Even though the judges explicitly admitted that these figures were 'not in [themselves] conclusive', the data were in any event capable of showing that, in spite of the wolf hunting, the conservation status of Grey wolves substantially and consistently improved in Finland during the pre-litigation stage.[73] This led the CJEU to conclude that the Commission had not adduced sufficient evidence as to the existence of an administrative practice by the Finnish authorities that was incompatible with the Habitats Directive. At first sight, this outcome seems counterintuitive, especially in view of the precautionary principle, which is often at the heart of the EU's case law. Should nature not have the benefit of the doubt? Have we not learned that the EU courts are prone to interpreting protection schemes progressively?

However, the CJEU's decision in the more recent *French Hamster* case puts the CJEU's take on science in infringement proceedings more into perspective when it comes to species recovery.[74] In its 2011 decision on wild hamsters (*Cricetus cricetus*) in France, the CJEU did not hesitate to underline the restoration potential which was referred to in Article 12(1) of the Habitats Directive.[75] In addition to implicitly stating that Article 12(1) of the Habitats Directive is to be read as an 'obligation of result', the CJEU reiterated the progressive stance of Advocate General Kokott, who had previously noted that 'an unfavourable conservation status gives rise to more far-reaching obligations for EU Member States (…) because the system of protection is intended to restore a favourable conservation status'.[76] Hence the importance of scientific reports on the population dynamics and the viability of the rodent species in France. When reviewing the evidence put forward by the European Commission in the *French Hamster* case, the CJEU held that several factors rendered the long-term survival of the species precarious at best. In particular, the CJEU underlined that 'there were no populations of the [European hamster] (…) which reached its *minimum viable population threshold*, which is estimated at 1,500 individuals spread over an area of contiguous suitable land of 600 hectares'.[77]

It is tempting to set aside the latter judicial decision as a relatively straightforward application of the prevailing science with respect to the concept of FCS. For one, France did not go to great lengths to question the validity of the scientific reports used by the European Commission to underpin its claims. The reports were drafted by the French administration themselves and thus questioning their objectivity was not an option. However, such a conclusion might obscure the myriad of complex scientific questions that still remain in this respect. Take for instance the CJEU's reliance on the concept of 'minimum viable population' (MVP). At first glance, invoking this concept when evaluating the FCS of the wild hamster makes perfect sense since it is a popular concept to measure the survival chances of endangered species. In 1981, Shaffer defined the concept of MVP as 'the smallest isolated population having a 99% chance of remaining extant for 1,000 years despite the foreseeable effects of demographic, environmental and genetic stochasticity, and natural catastrophes'.[78] However, it is generally accepted that MVP can be determined in numerous ways.

As of today, one of the most prominent methods to establish MVPs is the so-called population viability analysis (PVA), which uses demographic and environmental information to project future population dynamics.[79] Another method to estimate MVPs consists in determining the minimum area that a population needs to inhabit in order to escape environmental catastrophes. In recent years, however, a relatively large amount of attention has been paid to the evolutionary potential of a population (evolutionary MVP), being the population size required at equilibrium to balance the loss of quantitative genetic variation with the gain from mutation.[80] In view of the sharp decline of the genetic health of the remaining populations of hamsters in Western Europe, this obviously also needs to be taken into account when setting population targets. Recent research indeed finds that using well-defined breeding lines combined with a systematic reintroduction scheme is key to safeguarding the genetic viability of the few remaining hamster populations in Western Europe. To complicate matters even further, there is widespread agreement among scientists that for a population to be genetically viable, it must consist of at least 500 effective individuals, *i.e.* individuals who contribute to the genetic diversity of the population's offspring. As the effective population size is normally significantly less than the total population size, it is generally accepted that in general a total population threshold of 5,000 individuals will be required to ensure genetic viability.[81]

What scientific approach is to prevail when applying the Habitats Directive? For now, the CJEU has not handed down a clear-cut decision in which more substantial guidelines are provided regarding the concrete interpretation to be used in the context of the concept of favourable conservation status. Nor has it clarified to what extent account needs to be taken of other societal interests when opting for ambitious population targets. If the more progressive interpretation of the concept of FCS ultimately gained the upper hand in courtrooms, this would evidently have serious implications for the ambition level of many national conservation policies. If the genetics of a species are a determining factor when setting viable population targets, Member States are clearly required to direct their conservation efforts in respect of endangered Annex IV species beyond merely preventing extinction in the short term.

The latter observations are not merely of a theoretical nature. The aforementioned controversy regarding the reference state to be used when establishing population targets for the Grey wolf in Sweden is instructive. According to the European Commission, the Swedish wolf population, despite its increase during the past decade, has not reached a FCS. In its Reasoned Opinion, the European Commission referred to a number of studies showing that the problem with the wolf in Sweden is linked to genetic diversity.[82] Yet these findings were openly negated by the Swedish Supreme Administrative Court in its 2016 decision on the Swedish wolf hunting permits. Admittedly, the Swedish judges referred to the scientific reports that were produced by the Swedish nature conservation agencies. However, no direct reference was made to the criticism voiced by leading scientists such as Guillaume Chapron, who questioned the lack of genetic viability of the Swedish wolf population due to the limited input of new 'migrant' blood into the isolated wolf population (see *supra*). The Swedish judges took great care in accepting the adequacy of the scientific research presented by the Swedish government, even when the same scientific research was not sufficient to sway the European Commission.[83]

That is not to say that the reports produced by the Swedish government were manifestly flawed. The Swedish ruling merely illustrates the limits of science-based policy-making, thereby, possibly inadvertently, reasserting Kuhn's viewpoint regarding the impossibility of achieving objectivity through science. It turns out the same 'science' can be used both to validate and challenge existing conservation policies. Either way, it is crystal clear that the rationale present in this ruling is deferential. Instead of opting for a stringent substantive scrutinising of the evidence presented to it, the Swedish court clearly opted for a marginal check. Even when faced with the argument that during the years preceding the grant of the hunting permits no new migratory wolves had come to reinforce the Swedish wolf population, which indicated that the risk of inbreeding could soon manifest itself, the Swedish judges did not find it necessary to further question the scientific reports presented to them.

It might be tempting to assume that the above-mentioned controversy is less relevant for the implementation of the Natura 2000 site protection rules since such rules are linked to designated areas. And thus there appears to be more certainty when it comes to the concrete territorial scope of the applicable protection regimes. Yet even in the context of protected sites, scientific views on the ecological baselines to be used as yardsticks when authorising activities might differ. For one, in its landmark ruling in the *Waddenzee* case, the CJEU unequivocally underlined the need to use site-specific conservation objectives as a clear-cut and defining yardstick when applying Article 6(3) of the Habitats Directive to permit applications:

> So, where such a plan or project has an effect on that site but is not likely to undermine its conservation objectives, it cannot be considered likely to have a significant effect on the site concerned. Conversely, where such a plan or project is likely to undermine the conservation objectives of the site concerned, it must necessarily be considered likely to have a significant effect on the site.[84]

And thus one might quickly conclude that the CJEU clearly opted for a science-based approach to the decision-making procedure in the context of potentially harmful activities for Natura 2000 sites. Yet the *Waddenzee* rationale does not offer us more guidance as to the substantive underpinning of the site-specific conservation objectives. It only provides indirect evidence of the importance of science-based information when granting permits. Therefore the question remains whether the conservation objectives are to be regarded as mere scientific benchmarks when authorising harmful activities. The jurisprudential reference to the 'best available knowledge',[85] which is to be considered when gauging the effects of plans or projects within the framework of Article 6(3) of the Habitats Directive, might indicate that the answer to the question above is positive. However, this does not mean that other considerations of a more economic or social nature are completely off-bounds or irrelevant. For instance, in its 2008 ruling in the *Commission v France* case, the CJEU recognised that 'determining the conservation and restoration objectives in the context of Natura 2000 may require (...) the reconciliation of various conflicting objectives'.[86] This decision seems to give rise to conflicting lines of interpretation. Does it allow for the inclusion of arguments other than purely science-based elements, even when this entails that the area will no longer be able to contribute to the overall Natura 2000 network?

More recent jurisprudence reveals that, when the available ecological science indicates that the site is in an unfavourable conservation status, there is only a limited margin to revisit conservation objectives in view of economic arguments. Ensuring that the site concerned is able to contribute to the overall FCS at national level appears to be the default option. However, only when the first *Sweetman* case – a preliminary reference from the Supreme Court of Ireland – had to be addressed was the CJEU provided with the opportunity to give more guidance regarding the concrete substantive implications of the concept of FCS.

When confronted with the disappearance of a relatively small portion of a priority habitat – 1.47 ha of limestone pavement in a sub-area of a Natura 2000 site, which sub-area hosted approximately 85 ha of that very same habitat type – the CJEU reasserted the importance of the concept of FCS in its 2013 decision. According to the CJEU, the fact that the impacts were relatively minor when considered against the total area of the said habitat type was not important. This rationale clearly stands in the way of a 'death by a thousand cuts' scenario. To be more precise, the CJEU explicitly stated that Natura 2000 sites need to be preserved at a favourable conservation status. According to the CJEU, this entailed the lasting preservation of the constitutive characteristics of the site concerned that are connected to the presence of a natural habitat type the preservation of which was the objective justifying the designation of that site.[87] Interestingly, the CJEU determined that since the site concerned was designated as a Natura 2000 site because of the presence at that site of limestone pavement, a natural resource which, once destroyed, cannot be replaced, 'the conservation objective thus corresponds to maintenance at a favourable conservation status of that site's constitutive characteristics, namely the presence of limestone pavement'[88]. In other words, the CJEU seems to link the conservation objectives to the designation process, which is, as we have learned, to be viewed as an exclusively scientific enterprise (see *supra*). Henceforth, there appears to be no leeway to come forward with less ambitious conservation objectives.

A similar approach prevailed in the 2018 *Grace and Sweetman* case, which revolved around the negative impacts of a wind farm project at an Irish Natura 2000 site of great importance for the hen harrier (*Circus cyaneus*), a species of raptor. Here, although the CJEU noted that the

> designation of a territory as an SPA for the conservation of [hen harrier] entails the lasting preservation of the constitutive characteristics of the habitat in that area, the survival of the species in question and its reproduction being the objective justifying the designation of that area

the Court did not provide further guidance on the concrete determination of the said conservation objectives.[89] Likewise, in its landmark 2018 decision on the Dutch programmatic approach to the regulation of nitrogen deposition from agricultural activities, the CJEU held that

> (i)n circumstances such as those at issue in the main proceedings, where the conservation status of a natural habitat is unfavourable, the possibility of authorising activities which may subsequently affect the ecological situation of the sites concerned seems necessarily limited.[90]

In this case, the CJEU was faced with Dutch Natura 2000 sites that were overloaded with an excessive amount of nitrogen and the quest, nevertheless, for more leeway when granting permits for new agricultural developments. Further, in the *Leenheerpolder* case (discussed above), Advocate General Kokott clearly underlined the restoration obligations incumbent on Member States when confronted with Natura 2000 sites which are in an unfavourable conservation status,[91] while also underlining that political, social and budgetary reasons cannot constitute proper motivation for the reduction in size of a Natura 2000 site.[92] The main thread of the recent developments in case law appears to be that only manifest scientific errors can give rise to a revision of designated sites. Or, put differently, the available science should normally lead to more reliable baseline scenarios, which should in turn lead to sounder and frequently more stringent permitting procedures.

Science-based authorisation procedures: the precautionary principle as a catalyst for more sustainable conservation policies?

Science is not static. By definition, scientific progress is unpredictable and can at times even be erratic. And thus the precautionary principle, which is also reflected in Article 191 of the Treaty on the Functioning of the European Union (EU), inevitably comes to the forefront when discussing the interface between science and the Habitats Directive, especially within the context of the Natura 2000 network. As a legal principle, precaution is hotly debated. Uncertainty is inherent to human life and by opting for a too rigid understanding of 'precaution', one might ultimately block the emergence of sustainable innovations.[93] That said, the prominence of the precautionary principle is logical in a society which is depicted as a 'risk society'.[94] The latter is certainly increasingly visible in the context of environmental degradation, which is further exacerbated by phenomena such as climate change. And thus the principle is becoming increasingly relevant in the context of the Habitats Directive, especially in the context of permitting policies with respect to Natura 2000 sites.

This has much to do with the strict wording of Article 6(3) of the Habitats Directive: by setting strict substantive and procedural requirements to be followed with respect to a plan or project which is not directly connected with or necessary to the management of a Natura 2000 site but which is likely to have a significant effect thereon, Article 6(3) and (4) seek to pre-empt damage being done to the site or to minimise that damage.

This begs the question to what extent judges are required to carry out a substantive review of the scientific reports that are put forward by parties in the courtroom. Is the lack of scientific consensus on the adverse effects of certain projects sufficient to reject a permit application? Does the precautionary principle only come into play when the threatened harm is considered to be significant, excluding minor or trivial risks? Is a mere likelihood of effects sufficient for a permit refusal? And, ultimately, is there some margin of appreciation for courts to balance environmental, social and economic interests, under the general umbrella of the proportionality principle?

In this respect, the landmark ruling of the CJEU in the aforementioned *Waddenzee* case still stands out as the default option. As is widely known, the CJEU firmly held that the authorisation criterion laid down in the second sentence of Article 6(3) of the Habitats Directive integrates the precautionary principle. Hence competent national authorities are only permitted to allow projects or plans if they have made certain, in light of the appropriate assessment and the applicable conservation objectives, that the project or plan will not adversely affect the integrity of the Natura 2000 site(s) in question. That is the case where no *reasonable* scientific doubt remains as to the absence of such effects.[95] By placing the burden of proof on the proponent of the potentially harmful activity, the CJEU opts for the 'prohibitive' formulation of the precautionary principle in the context of Natura 2000. In the end, it will be for the proponent of an activity to put forward the necessary conclusive evidence as regards the absence of adverse effects in order to enable the permitting authority to ascertain that the plan or project would not give rise to adverse effects at a Natura 2000 site.[96]

Most prominent in this respect is the CJEU's emphasis on the importance of using the best scientific knowledge when carrying out a so-called 'appropriate assessment' in order to enable the competent authorities to conclude with scientific certainty that there will be no adverse effects on the site's integrity.[97] In other, more recent, case law the CJEU also reiterated the importance of assessments that do not confine themselves to a mere repetition of the existing data available regarding certain sites. Accordingly, an appropriate assessment should contain complete, precise and definitive findings and conclusions capable of removing all reasonable scientific doubt as to the effects of the proposed works.[98] In other words, when such assessments contain manifest lacunae or inadequacies, or even outright errors, they cannot be used as a basis for granting a permit pursuant to Article 6(3). Of course, the relatively stringent requirement linked to definitive findings does not automatically entail that an appropriate assessment needs to exclude all potential risks. The natural environment is uncertain by nature. The certainty requirement cannot be construed as meaning an absolute, ontological certainty since this would give rise to disproportionate effects.[99]

As such, while it is perfectly possible to raise philosophical doubts in every single decision-making process, this would lead to inertia. Seeing Kuhn's interpretation of science as a flexible, subjective and multi-dimensional process, it comes as no surprise that a more static legal approach would seem to give rise to an obstacle course for those seeking and granting permissions. And thus courts will be asked to carry out a balancing exercise in this respect, distinguishing between pertinent scientific uncertainty and superfluous arguments. On the one hand, project developers can perhaps not be expected to carry out new scientific research in order to bolster their arguments in view of climate change when the available climate science is not yet conclusive when it comes to the shifting of habitats as a result of climate change. On the other hand, Article 6(3) requires conclusive evidence regarding the absence of effects. Much will depend on the precise nature of the plan or project in question. For instance, more certainty will be required for activities that might directly lead to increased CO_2 emissions, since such activities will directly contribute to global heating. And even when assessing the long-term impacts on Natura 2000 sites of activities that do not directly emit CO_2, climate science will play a crucial role in determining the future baseline against which the significance of plans and/or projects may be further assessed. For instance, in light of the increased chances of periods of sustained drought in western Europe, even the mere continuation of existing groundwater abstraction activities might become problematic for vulnerable water-dependent habitats. In other words, climate projections will have to be integrated into the conservation objectives for Natura 2000 sites in order to allow for a comprehensive review of the acceptability of new and ongoing plans and/or projects.

The above-mentioned Polish *Białowieża* case is perhaps the most indicative example as to how scientific controversy can be used as a lever for better environmental protection within the context of Article 6(3) of the Habitats Directive. In this context, the European Commission was of the opinion that the felling of trees, mainly spruces, affected by the spruce bark beetle, and the removal of dead and dying trees, deadwood and centuries-old forest stands, flouted many rules enshrined in the EU Nature Directives. As indicated in the introduction to this chapter, Poland contested the scientific reports cited by the European Commission. This scientific controversy had to be addressed by the CJEU when addressing the Commission's request for interim measures. However, according to the CJEU, 'the procedure for interim relief is not designed to establish the truth of complex facts that are very much in dispute'.[100] The CJEU also pointed out that when hearing an application for interim measures it must postulate, solely for the purpose of the assessment of the existence of serious and irreparable damage, that the complaints put forward in the main proceedings by the applicant for interim measures might be upheld.[101] As a consequence, the simple fact that the scientific opinions referred to by the parties are contradictory cannot in itself lead to the dismissal of the request for interim measures. In fact, the CJEU used both the *Waddenzee* rationale – i.e. the reference to best available scientific knowledge – and the general precautionary principle as an argument to conclude that the European Commission had a prima facie case.[102]

Rather than getting caught up in the heated scientific debate regarding the potential effects of the logging activities, the CJEU opted for a more pragmatic, common-sense approach. The court did not resort to appointing experts to assess the value of the conflicting scientific reports, which is possible pursuant to the procedural rules of the CJEU but would lead to valuable time delays (see *infra*). The main focus was instead on the scientific elements and facts that had remained *uncontested* between both parties. In essence, the Polish authorities had not challenged the premise that the forest management actions would have a negative impact on the relevant habitats, which mainly consisted of old, dying and dead trees. This was also indirectly taken into account in the purported management plans, which left some areas untouched precisely because of the presence of valuable natural habitats.[103] Ultimately, the CJEU drew on these findings to conclude that there was a

> prima facie lack of scientific information excluding beyond all reasonable doubt that the active forest management operations at issue have damaging and irreversible effects on the protected habitats of the Natura 2000 Puszcza Białowieska site referred to in the Commission's action.[104]

This remarkably progressive approach to the use of conflicting scientific reports in the context of proceedings for interim measures could be replicated in national court cases, especially since it demonstrates that the simple presence of conflicting scientific studies can by no means be used as an argument to deny requests for interim measures regarding the protection of endangered habitats and species awaiting a final ruling on the merits. Clearly, this line of interpretation allows for more effective judicial protection in the field of EU nature conservation, preventing fait accompli scenarios from arising.

A similar stringent approach also prevailed in the context of the CJEU's ruling on the substance of the case. Again, no experts were appointed by the CJEU. In reaching its final decision, however, the CJEU did not refrain from a more in-depth analysis, which prompted it to review the substantive underpinnings of the Polish forest management programmes.[105] For instance, in paragraphs 173 and 174 of the ruling, the CJEU noted that the spruce bark beetle was not listed as a direct threat to the Białowieża forest prior to 2015, nor was there a clear link between the volume of wood harvested and the trees that were directly impacted by the beetle. These inconsistencies did not permit a finding that the active forest management operations at issue may be justified by the need to stop the spread of a harmful organism of that kind. Again, the CJEU also referred to the precautionary principle as a principal yardstick, and held that:

> it is apparent from the material supplied to the Court and the exchange of arguments at the hearing that, as at the date of adoption of the 2016 appendix, there was still scientific controversy regarding the most appropriate methods to stop the spread of the spruce bark beetle. As is clear from the remediation programme, that controversy related, in particular, to the very desirability of combating its spread, which, according to certain views in the scientific community, formed part of a natural cycle corresponding to periodic trends specific to the characteristics of the site whose preservation was the objective justifying the site's designation in the list of SCIs and as an SPA. Consequently, (...), the Polish authorities could not adopt the 2016 appendix, as there was no scientific certainty that the active forest management operations at issue would not have lasting adverse effects on the integrity of the site concerned.[106]

As highlighted above, it would be incorrect to deduce from the latter case that simple scientific controversy would be sufficient to ban certain harmful activities in a Natura 2000 context, as this could lead to repercussions even for sustainable projects. Even so, when the controversy in question touches upon the fundamental impacts that are at stake in the context of an ecological evaluation for a potentially harmful plan or project, one should infer from the latter jurisprudence that authorisation can no longer be the default option. It serves as a clear recognition of the increased power of sound science when contesting economic activities prone to impact EU-protected nature. And, most importantly, it gives national courts a pragmatic pathway to deal with conflicting scientific reports without having to have recourse to the appointment of scientific experts, which will often give rise to additional delays and disputes.

However, during the past decade, a remarkable novel twist was given by the CJEU regarding the use of science in the context of the integration of mitigation actions in an appropriate assessment for a plan or project. This was most apparent in the field of habitat and ecological restoration, which was increasingly relied upon as an expedient mitigation measure to allow for the authorisation of a plan or project without having recourse to the derogation clause in Article 6(4) of the Habitats Directive. In many instances, Member States integrated restoration actions into planning permits as a means of counterbalancing the inherent negative effects linked to project developments.[107] This in spite of the fact that the science underpinning the use of future habitat restoration as an immediate offset for project development remained ambivalent, to say the least. An increasing number of authors have claimed that offset policies are not evidence-based.[108] And thus it was not surprising to see that the CJEU opted for a restrictive approach. In its 2014 decision in the *Briels* case, the CJEU held that restoration actions could not, as a matter of principle, be considered preventative or mitigation measures in the context of an appropriate assessment if the purpose was to offset actual damage to protected habitats.[109] This is common sense: granting a permit for nature destruction with reference to offsets at another location can hardly be branded a 'mitigating action'. Most importantly, though, the CJEU also relied on the precautionary principle:

> It should further be noted that, as a rule, any positive effects of a future creation of a new habitat which is aimed at compensating for the loss of area and quality of that same habitat type on a protected site, even where the new area will be bigger and of higher quality, are highly difficult to forecast with any degree of certainty and, in any event, will be visible only several years into the future, a point made in paragraph 87 of the order for reference. Consequently, they cannot be taken into account at the procedural stage provided for in Article 6(3) of the Habitats Directive.[110]

In the CJEU's opinion, one cannot anticipate future recovery actions in an appropriate assessment. There needs to be scientific certainty regarding the absence of impacts at the time of the evaluation of the effects. Put differently, the CJEU decided to consistently reapply its *Waddenzee* rationale to every single measure or action, including mitigation or compensation actions, linked to a project development. This approach was subsequently refined by the CJEU in the *Orleans* case, where the CJEU rejected a proactive compensation scheme that had been used as an argument to justify future harbour development leading to the irreversible destruction of protected natural habitats. The CJEU reasoned that Article 6(3) of the Habitats Directive made impermissible an approach under which the effects of future restoration measures, which were still uncertain at the time of the appropriate assessment, are already accounted for in the authorisation.[111] In paragraph 55 of the decision, the CJEU reiterated the importance of sound data and sufficiently tight timelines when implementing restoration actions. It held that the

> benefits resulting from the creation of the nature reserves have already been taken into account in the assessment and in demonstrating the absence of significant adverse effects on the site even though the result of the creation of those reserves is uncertain, since it is not complete.

In the above-mentioned *Grace and Sweetman* case, a similar discussion emerged regarding the use of restoration actions as a means of bypassing the application of the strict derogation clause in Article 6(4) of the Habitats Directive. Once more, the uncertain nature of the actions proved to be the Achilles heel of the challenged authorisation. Flexible management actions aimed at the recovery of habitats for the hen harrier were to serve as a mitigation measure for the adverse impacts linked to a wind farm. The CJEU, however, explicitly underlined in paragraph 53 of its decision that the mere qualification of the actions as dynamic management was not the main reason for the dismissal:

> It is not the fact that the habitat concerned in the main proceedings is in constant flux and that that area requires 'dynamic' management that is the

> cause of uncertainty. In fact, such uncertainty is the result of the identification of adverse effects, certain or potential, on the integrity of the area concerned as a habitat and foraging area and, therefore, on one of the constitutive characteristics of that area, and of the inclusion in the assessment of the implications of future benefits to be derived from the adoption of measures which, at the time that assessment is made, are only potential, as the measures have not yet been implemented. Accordingly, and subject to verifications to be carried out by the referring court, it was not possible for those benefits to be foreseen with the requisite degree of certainty when the authorities approved the contested development.[112]

In spite of this nuance, the CJEU's recent case law remains critical as to the use of vaguely described adaptive management techniques in the context of Article 6(3). For instance, in the *Commission v Germany* case, the CJEU further underlined the need for definitive data, a requirement which cannot simply be circumvented by including a monitoring protocol in the permit.[113]

Most prominently, the limited potential of liberal adaptive management approaches came to the fore in the CJEU's 2018 decision on the compatibility of the Dutch programmatic approach to nitrogen deposition with Article 6(3) of the Habitats Directive.[114] This programmatic approach encompassed a general, flexible and adaptive approach to nitrogen deposition in the agricultural sector, combining both future reduction and restoration actions with economic developments. The approach was rooted in comprehensive scientific studies regarding the ecological effects of nitrogen deposition and online calculating tools.[115] Nevertheless, the approach did not survive judicial review. The CJEU did not explicitly rule out the performance of an assessment at the level of a programme, especially since it might allow for a better appraisal of possible cumulative effects.[116] Even so, it tasked the national court with the duty to carry out 'a thorough and in-depth examination of the scientific soundness of that assessment' in order to ascertain whether no reasonable scientific doubt remains as to the absence of such effects in the context of an adaptive approach to nitrogen deposition.[117] This undoubtedly represents a high hurdle and begs the question of whether a high-level assessment will ever be capable of providing a sufficient level of certainty. Especially since the CJEU equally highlighted in its decision that future benefits of restoration measures that have not been implemented cannot be taken into account in an appropriate assessment precisely because 'the level of scientific knowledge does not allow them to be identified or quantified with certainty'.[118]

This string of rulings adds a new layer to the jurisprudence regarding the precautionary principle and indicates that the CJEU is clearly setting a high water mark when it comes to the use of restoration actions in the context of permitting regimes. The CJEU did not in any of the cases explicitly quote or refer to specific scientific studies in support of the limited effectiveness of the said habitat restoration actions. For instance, there is ample scientific research that reaffirms the inadequacy of restoration measures in the context of natural habitats that have been overloaded with nitrogen.[119] These studies and reports were indeed even included in the pleadings of the opponents of the Dutch programmatic approach to nitrogen. Equally the CJEU did not seek the advice of technical experts. Be that as it may, in light of its reassertion of the precautionary principle it appears virtually impossible to take into account the future effects of restoration or even other mitigation measures that have not been accounted for in an appropriate assessment. Only when such actions have been scientifically tested before and proven to be effective may it be possible to take them into account before they have been implemented.

The CJEU's strict stance might be used by opponents of the Habitats Directive as an illustration of the Directive's alleged rigidity. The recent case law indeed underlines the importance of conclusive scientific evidence when including novel management techniques in permitting decisions. In the absence of such science, permitting procedures will turn into obstacle courses or, depending on one's perspective, effective instruments to avoid further degradation.

Feelings vs science: dealing with large carnivores

Science is also becoming increasingly relevant in the management of large carnivores. Benefiting from the extra layer of protection offered by the Habitats Directive, species such as the Grey wolf have incrementally recolonised former areas in which they once thrived. However, the growing populations of large carnivores such as these have in practice caused concern, especially among livestock owners and hunters. One way of dealing with this is to allow the culling of wolves based on the idea that this will increase tolerance to the wolves' presence and decrease the number of conflicts between large carnivores and livestock keepers. This is what I referred to as 'tolerance hunting' above, a technique that has gained a lot of traction in Scandinavia.[120] However, the scientific underpinning of this policy shift appears uncertain. A 2012 longitudinal study focusing on the impact of the legalisation of wolf hunting in Wisconsin revealed that the attitude towards wolves continued to decline, even after the wolf was delisted as a protected species in this U.S. state.[121] Predation rates also have no linear relationship with the abundance of predators but are often more closely linked to subtle factors such as their complex social system and social behaviour and population dynamics.[122]

In light of the lack of conclusive scientific evidence underpinning wolf-hunting strategies, it is interesting to take a quick glance at the CJEU's track record in this respect. In its above-mentioned 2007 decision on hunting strategies in Finland aimed at reducing the impact on livestock, Advocate General Kokott explicitly relied on scientific studies when discussing the fact that Finland refused to restrict its shooting permits to particular animals:

> However, Finland's above submissions do not specify how shooting permits which are not specific to individual wolves contribute to preventing

> serious damage, that is to say whether this practice is appropriate at all for attaining this objective. In North America, wolf populations had to be *greatly* reduced over a fairly long period of time before losses of game decreased. It cannot be ruled out that this also applies to harm to livestock.[123]

It is exceptional to see an Advocate General *proprio motu* bring forward scientific studies in the context of an infringement procedure. In its subsequent ruling, the CJEU reasserted the Advocate General's opinion, stating that

> certain parties are of the opinion that continued hunting keeps wolves wary of humans and thus helps to reduce damage, while others consider that hunting of wolves which belong to packs only increases damage. Furthermore, it is stated that little biological research on this topic is available.[124]

Given the lack of scientific consensus and the impossibility of targeting individual wolves in the Finnish wolf strategy, the CJEU concluded that Finland had not proven that its hunting practices were preventive, as required by Article 16(1)(b) of the Habitats Directive.

In view of this setback, some Scandinavian countries, such as Sweden and Finland, refrained from invoking Article 16(1)(b) and tried to accommodate tolerance hunting strategies by means of Article 16(1)(e) of the Habitats Directive. This arguably enigmatically formulated derogation provision grants Member States the possibility 'to allow, under strictly supervised conditions, on a selective basis and to a limited extent, the taking or keeping of certain specimens (...) in limited numbers specified by the competent national authorities'. As highlighted above, the Swedish Supreme Court accepted the limited scientific evidence underpinning tolerance hunting in the case before it, stating that it had no reason to question the county board's finding that hunting can increase societal acceptance of the presence of large carnivores. Yet, in its 2019 decision on the compatibility of Finnish tolerance hunting practices with Article 16(1)(e) of the Habitats Directive, the CJEU did not replicate the lenient view of the Swedish Supreme Administrative Court. On the contrary, the CJEU did not hesitate to put science back at the heart of the debate by urging national courts to scrutinise the scientific underpinnings of tolerance hunting.[125] Whereas it did not outright reject tolerance hunting as being incompatible with Article 16(1) of the Habitats Directive, the list of conditions to be observed is so long that it remains doubtful whether such practices might ever be compatible with the strict conditions of Article 16(1)(e) of the Habitats Directive. The CJEU held that

> it is for the national authority to support, on the basis of rigorous scientific data, including, where appropriate, comparative data on the effects of hunting for population management purposes on the conservation status of wolves, the proposition that hunting for population management purposes

> is actually capable of reducing illegal hunting to such an extent that it would have a net positive effect on the conservation status of the wolf population, whilst taking account of the number of derogation permits envisaged and the most recent estimates of the number of wolves taken illegally.[126]

Additionally, the CJEU underlined the need for 'the best relevant scientific and technical evidence' to prove that the objective pursued by tolerance hunting cannot be achieved by other means, such as a more effective enforcement of the Habitats Directive.[127] As in other cases, the CJEU also reiterated the importance of the precautionary principle.[128] Adding to the other stringent conditions, the CJEU stated that if

> after examining the best scientific data available, significant doubt remains as to whether or not a derogation will be detrimental to the maintenance or restoration of populations of an endangered species at a favourable conservation status, the Member State must refrain from granting or implementing that derogation.[129]

Again, we see a clear willingness on the part of the CJEU to fully engage with the need for science-based decision-making in the field of species protection. In doing so, the CJEU aligns its jurisprudence regarding strictly protected species with the protection of Natura 2000. Yet the effects of the CJEU's move are not to be underestimated, as politicians are often inclined to give in to vocal minorities aimed at safeguarding the status quo when it comes to recovery policies. Beyond that, it seems clear that the CJEU has strategically opted to regard science-based policy-making as the baseline when reviewing conservation policies. For instance, instead of categorically rejecting tolerance hunting in a context where a species finds itself in an unfavourable conservation status, the CJEU compelled the national authorities to provide the courts with sufficient and complete scientific data in support of their claims. Possibly the CJEU was aware of the lack of scientific support for tolerance hunting. Yet, instead of choosing sides, the CJEU skilfully avoided becoming fully engaged in this scientific debate by simply urging the national authorities to produce complete and definitive data supporting the hunting derogations. While a lot can be said in support of the CJEU's pragmatic stance, this means that an inherent weakness is surfacing which cannot remain unnoticed. Indeed, the increasing reliance on 'science-based' procedures does not answer the more fundamental question: what is to be treated as sound, objective science? What types of scientific research fall within the notion of 'science' that is considered by the CJEU? Is this confined to population-based research on endangered species or does social science also clear the hurdle set by the CJEU? It remains to be seen whether the CJEU will be found ready to further clarify these questions in future cases.

A wavering court?

The analysis presented above gives rise to a certain paradox. On the one hand, the CJEU can be seen to be increasingly ready to underline the science-based nature of the decision-making procedures required in the context of EU-protected habitats and species. However, at the same time, the CJEU is particularly cautious not to get caught up in a fierce battle of expertise between proponents and opponents of certain project developments or activities liable to have adverse effects on the environment. In this respect, the position of the CJEU is not that much different from that of the ICJ: both appear reluctant to use so-called expert evidence.[130] What's more, in none of the environmental cases analysed above did the CJEU find it necessary to commission expert opinions.

Even so, pursuant to Article 25 of the CJEU statute, '(t)he Court may at any time entrust any individual, body, authority, committee or other organisation it chooses with the task of giving an expert opinion'. Such experts can be commissioned not only in the context of infringement actions brought by the European Commission but also in preliminary reference proceedings. However, as noted above, even in the context of highly technical and debated issues, such as the Dutch programmatic approach to nitrogen deposition or tolerance hunting, the CJEU did not find it necessary to ask for expert opinions. Equally, the appointment of experts was not requested by the parties to the case. This finding is by no means exceptional: generally speaking, the CJEU also appears very reluctant to call upon expert opinions. The reasons for this are likely to be numerous, ranging from the specific function of the CJEU as interpreter of EU law to the reliance by the EU judges on their own expertise and the strong inclination to retain as much control as possible over the adjudication of purely technical matters.[131] This outcome may appear to be counterintuitive since environmental law is often characterised as a field of law known for its elevated level of technicality. As of today, however, it is clear that the CJEU's deference is inextricably linked to matters such as its application of the burden of proof, especially in the context of infringement cases, as well as the precautionary principle. Moreover, when asked to interpret EU law in the context of preliminary reference proceedings, it is not up to the CJEU to resolve disputes relating to the factual elements at play in the case at hand.[132] The CJEU's role does not consist in certifying the validity of the facts or even science brought forward by the parties. Especially in the context of preliminary reference proceedings, the main objective is to provide the national court with a sound and (relatively) quick interpretation of EU law. The appointment of experts before the CJEU might further complicate matters.

The prevalence of references to the precautionary principle obviously helps to explain this more reserved approach. But there is more. The CJEU appears to see little added value in appointing experts even when such experts might pinpoint manifest defects in scientific studies. One might indeed wonder whether the intervention of technical experts might ultimately serve to solve at least some of the scientific questions at play in court proceedings. Will appointed experts be able to solve technical discussions that are not yet resolved in the academic literature? National courts might enjoy less flexibility in this respect. As national courts are required to provide effective judicial protection in environmental cases, they will also be required to carry out a substantive review of the facts presented before them. They might thus be more eager or willing to rely on the input of technical experts. Even so, national courts often avoid using experts when possible. For instance, in its final decision on the Dutch programmatic approach to nitrogen deposition, the Dutch Council of State found it unnecessary to appoint further experts in order to review the evidence regarding the potential impact of elevated levels of nitrogen deposition on Dutch Natura 2000 sites. It simply used the conflicting scientific outputs, the lack of implementation of the recovery actions prior to the permit procedure and the unclear relationship between existing conservation duties and the compensatory actions as arguments to quash the Dutch programmatic approach to nitrogen deposition.[133]

New ways forward: entering a brave new world?

Confronted with the ever more important interactions between science and the Habitats Directive in practice, the question remains how we should accommodate ourselves as lawyers, policy-makers, NGOs and business people to this new reality. The assumption of science-based decision-making procedures – which is very much present within the recent case law developments at EU level – unavoidably affects future policy developments and litigation strategies but also urges scientists and citizens to rethink their traditional roles, beyond the sharp dichotomy between the fields of science and activism. While it remains hard to draw definitive lessons, the elements listed below will inevitably be at the forefront of future debates.

Adaptive management to the rescue

A logical reaction to the stringent case law of the CJEU regarding precaution and scientific uncertainties will be an increased reliance on adaptive management strategies.[134] Such strategies would allow permitting authorities to cope with the inherently uncertain nature of ecosystems. The inclusion of strict monitoring protocols allows agencies to revise existing permits in view of newly emerged scientific findings. For example, in its ruling on the Dutch programmatic approach to nitrogen deposition, the CJEU underscored the importance of integrated and programmatic approaches, especially since they allow the competent authorities to take into account cumulative effects. However, adaptive management is no panacea for all ills. More adaptive and flexible permitting strategies will only be compatible with Article 6 of the Habitats Directive if they are used to tackle the real environmental issues at stake. This was precisely lacking in the Dutch programmatic approach, which anticipated future recovery and deposition reduction actions as an (implicit) argument to justify the granting of additional permits for new economic developments.

Consequently, adaptive management strategies could certainly be identified by national authorities as the obvious answer to the increased scrutiny required by the CJEU, yet such strategies might only survive legal review where clear-cut commitments going beyond the existing ecological baseline are identifiable. Adaptive management strategies that merely serve as a cover for economic development against the backdrop of existing deteriorating environmental baselines or as an excuse not to meddle with unsustainable lock-ins for certain harmful activities will ultimately fail in the courtroom when the achievement of favourable conservation status is not sufficiently ensured.[135] Only when such strategies are aimed at effectively limiting existing environmental pressures might they be in line with the Habitats Directive.

A new role for scientists

Then there is the new role for scientists. For a considerable time, many scientists may have been inclined to believe that their research was distinctively separate from the political sphere of decision-making. And thus most scientists may have held the belief that their research would have no immediate impact on the discretion enjoyed by decision-makers when it came to deciding on unsustainable project developments. The increased insistence on science-based decision-making by the CJEU underlines the important impact of scientific research. Indeed, when judges are bound to substantively review the science underpinning decisions regarding EU endangered natural habitats or species, the value of scientific studies increases exponentially as an immediate result. For example, when scientists conclude that an EU-protected species is in an unfavourable conservation status in an EU Member State, this finding might have serious repercussions on the margin of discretion available to national authorities when it comes to considering the permissibility of new harmful activities.

Given the increased practical repercussions of scientific research in court, sufficient guarantees should exist to shield scientists from political pressure, which might be expected to increase. For one, as illustrated above, the growing impact of scientific research in litigation could lead to a reduction of public funding or other financial support for certain areas of scientific research. Or even worse, it could lead to outright interference with the outcome of scientific studies regarding EU-protected nature. Scientists should surely continue to apply high ethical standards when conducting their research. Yet, even more importantly, scientists should remain protected from political interference so as to ensure that they do not shy away from issuing clear-cut findings on key environmental threats.

Citizen science on the rise

In view of the increased review possibilities and the procedural rights granted to the general public through the Aarhus Convention, the importance of citizen science when implementing conservation strategies and permitting new adverse impacts has only increased. Citizen science is a general term for scientific research conducted, in whole or in part, by amateur (or non-professional) scientists.[136] It is described by some as 'public participation in scientific research', participatory monitoring or participatory action research, which underlines the inextricable links between citizen science and public participation. In times when people appear to be increasingly distrustful of scientific experts, citizen science might allow citizens to fully engage with professional scientists. This could in turn strengthen the interface between universities, scientific institutes and the wider public. In addition, recent research has demonstrated that citizen science has enormous potential to advance science. It has been revealed that citizen science projects can influence policies and guide resource management by producing datasets that are otherwise infeasible to generate.[137]

As a consequence, decision-makers can no longer ignore the potential of citizen science, especially in matters which directly impact the wider public. A failure to take into account the available science from the public might not only jeopardise the effectiveness of conservation strategies but also lead to legally flawed planning permits and juridical outcomes. Effective participation in decision-making procedures relating to the conservation of protected sites and species will be of seminal importance in the decades to come. This approach has already been reaffirmed by the Belgian Constitutional Court, which held that Article 7 of the Aarhus Convention requires consultation with the wider public when drafting management strategies for Natura 2000 sites.[138]

New litigation strategies

A final observation relates to the impact of science on litigation strategies. Up to now, most of the environmental litigation in the context of EU-protected habitats and species has been of a reactive nature: the main focus has been on challenging the legality of permits for potentially harmful plans and projects. This makes sense since new damaging development might lead to further deterioration of protected sites. Moreover, for a considerable time, direct legal challenges against overarching policy plans were deemed to be contrary to the separation of powers. As is apparent from the above analysis, the reassertion of the importance of basing permits on sound and complete scientific evidence as to the absence of potentially damaging impacts has made the 'reactive' route even more promising for environmental NGOs. Permits need to be science-based. However, it can be expected that the concrete operationalisation of the concept of 'favourable conservation status' will also gain traction in courtrooms, as the case law referred to above indicates that environmental NGOs have been granted the procedural right, on the basis of e.g. Article 9(3) of the Aarhus Convention, to seek judicial review of the scientific underpinnings of the general conservation strategies required to implement the conservation duties imposed by Articles 6 and 12 of the Habitats Directive. If environmental NGOs manage to demonstrate that the competent authorities in question have fallen short of their conservation duties, then it is no longer science fiction to envisage a national court ordering the competent authorities to step up their recovery actions. In principle, the burden of proof regarding the scientific soundness of the adopted conservation strategies lies with the competent authorities.

By directing future litigation towards Member States' broader conservation strategies, the piecemeal, reactive and defensive character of classic administrative annulment procedures might be overcome.[139] More proactive lawsuits are emerging, as is also illustrated in the context of climate change litigation, where the inadequacy of national mitigation policies is directly challenged in national courts. Such third-party enforcement procedures might also benefit the overall effectiveness of the Habitats Directive. A recent ruling of the administrative court of Toulouse, in which France was ordered to step up its recovery actions for Brown bears in the Pyrenees, serves as a powerful illustration of this new litigation trend. In reaching this conclusion, the court reasoned that the existing efforts were inadequate to reach FCS.[140]

Specialised environmental courts

In view of the increasingly scientific and technical character of existing procedures regarding the application of the Habitats Directive, there is arguably an increasingly urgent need for the establishment of specialised environmental courts. Not necessarily at EU level, given the sui generis role of the EU courts, but certainly at national level. In some countries, important steps have already been taken in this direction. This is for instance the case in Sweden, where the Swedish environmental code stipulates that each regional environmental court must have a panel consisting of one law-trained judge, one environmental technical advisor and two lay expert members.[141] The judge and the technical advisor are employed by the court and work as environmental judges full-time. Most importantly, all four members of the panel are equals in the decision-making process. It is important to highlight that the technical experts are often highly specialised academic scholars. However, they must also be familiar with the relevant procedural rules governing the cases at hand. This is but one example.

Of course, one should be wary of over-expectation. It has been noted above that the Swedish supreme administrative court's position on tolerance hunting contradicts the CJEU's view. Certainly, the presence of technical judges might be capable of overcoming the multitude of difficulties judges face when analysing the scientific evidence presented to them. However, this is certainly no panacea. In some instances, the added value of technical judges might remain limited. Courts might, for example, still prefer to rely on technical experts if they are unable to autonomously appraise the quality and objectivity of the scientific information provided to them. As demonstrated in this chapter, it is possible for judges to review decisions in view of the available science without having to defer to technical experts.

Conclusion: doing the impossible?

The above analysis aimed to outline the many questions at stake regarding the increasingly complex interface between science and law. Environmental court cases have become increasingly more technical and fact-based, which begs the question of whether traditional courts are still sufficiently equipped to offer effective judicial protection. The main thesis of this chapter is that the CJEU has drawn a clear line in the sand: both national and EU courts can no longer remain silent when it comes to substantive review in cases relating to the application of the Habitats Directive. In the context of both Natura 2000 and strict species protection, the CJEU has repeatedly underlined the importance of science-based decision-making procedures. Whenever inconclusive data are present, courts and judges are obliged to critically review the available science presented to them and provide effective judicial protection. The strict application of the precautionary principle urges courts to refuse permits for potentially harmful practices based on uncertain data, even without having to appoint technical experts to verify the conflicting data. However, this strict jurisprudence renders courts increasingly vulnerable to criticism, especially if they opt for lines of interpretation which lead to the tightening up of permitting policies for economic sectors like dairy farming or the coal industry. Concerns relating to *trias politica* might loom around the corner.

Even though it is tempting to applaud the CJEU's move to science-based decision-making, one should not remain blind to its inherent weaknesses. Science is always in flux. Science can be abused. The concrete examples referred to in this chapter, such as tolerance hunting in Scandinavia and Polish logging practices, clearly illustrate that science is prone to subjectivity. Feelings and facts get easily mixed. Or, to cite Thomas Kuhn, quoting Francis Bacon, 'Truth emerges more readily from error than from confusion.' Technical judges might be more equipped to distinguish emotional from scientific arguments. Yet science devoid of the subjective human experience does not exist. Therefore the paradoxical conclusion might be that environmental lawyers are asking judges and courts to do the impossible: to rule on the objectivity of scientific evidence presented to them. The linearity of science remains uncertain. Paradigms might shift. Furthermore, judges will ultimately not be able to escape the prevailing scientific frames. As of today, one can only hope that the judiciary is critical and competent enough to understand both the limits and the possibilities of adjudication in the realm of environmental litigation. It is of the utmost importance that judges learn to understand the process of science. This should be the overarching lesson drawn from this analysis. For the time being, the CJEU seems to have achieved precisely that objective. The question arises whether it will be able to maintain this stance in the coming years without the assistance of technical experts. And whether the Orwellian nightmarish vision of 'doublespeak' can be safely kept out of the courtroom.

Notes

1 George Orwell '1984' (1949, published by Penguin Books, 2008).
2 Council Directive 92/43/EEC on the conservation of natural habitats and of wild fauna and flora (1992) OJ L 206/7.
3 Thomas S. Kuhn, 'Objectivity, Value Judgment and Theory Choice', *The Essential Tension* (University of Chicago Press 1977), ch. 13. See also Thomas S. Kuhn, *The Structure of Scientific Revolutions* (University of Chicago Press 1962).
4 Quotation retrieved from: Erik Stokstad, 'Beetles Are Ravaging Europe's Oldest Forest. Is Logging the Answer?' (December 5, 2017) Sciencemag.org, available at https://www.sciencemag.org/news/2017/12/beetles-are-ravaging-europe-s-oldest-forest-logging-answer.
5 For more background, Valérie Dupont, Hendrik Schoukens and Charles-Hubert Born, 'Belgium', in Wolfgang Wende et al (eds.), *Biodiversity Offsets, European Perspectives on No Net Loss of Biodiversity and Ecosystem Services* (Springer 2018), 55.
6 See (in Dutch): 'Bioloog: 'Geen waardevolle bomen in het Essers-bos gevonden', (June 6, 2016) De Morgen.
7 In 2018, the Belgian Council of State quashed the permit because of its failure to take into account the cumulative impacts of an earlier expansion, which had to be offset in the area destined for the purported expansion. Belgian Council of State, no. 241.048 (March 20, 2018) Vzw Natuurpunt et al.
8 For more background, see Jan Darpö and Yaffa Epstein, 'Under Fire from All Directions. Swedish Wolf Management Hunting Scrutinized by Brussels and at Home', in Charles-Hubert Born et al (eds.) *The Habitats Directive in its EU Environmental Law Context. European Nature's Best Hope?* (Routledge 2016) 348.
9 Yaffa Epstein, 'Killing Wolves to Save Them? Legal Responses to 'Tolerance Hunting' in the European Union and United States', (2017) 226(1) RECIEL 19.
10 Guillaume Chapron, 'Challenge the Abuse of Science in Setting Policy' (2014) 516 *Nature* 289.
11 See for more background, Epstein (n 9).
12 Council Directive 79/409/EEC on the conservation of wild birds (1979) OJ L 103/1 (also referred to as 'Birds Directive'). The initial Birds Directive has been codified in European Parliament and Council Directive 2009/147/EC on the conservation of wild birds (2010) OJ L 20/7.
13 Aarhus Convention on access to information, public participation in decision-making and access to justice in environmental matters of 1998 (2005) OJ L 124/4.
14 Chapron (n 10).
15 Tracey Kanhanga, 'Scientific Uncertainties: A Nightmare for Environmental Adjudicators', in Christina Voigt (ed.), *International Judicial Practice on the Environment* (Cambridge University Press 2019), 121.
16 Ibid.
17 Alvin M. Weinberg, 'Science and Its Limits: The Regularo's Dilemma' (1985) 2(1) *Issues in Science and Technology* 60.
18 Marc Mangel, 'Whales, Science, and Scientific Whaling in the International Court of Justice,' (2016) 113(51) *Proceedings of the National Academy of Sciences* 14523–14527.
19 The analysis of the case law of the International Court of Justice draws on the more extensive review of the recent case law of the ICJ of Kanhanga (see n 15). This analysis is very helpful in our analysis of the jurisprudence of the Court of Justice of the EU when it comes to the use of science in the context of the Habitats Directive.
20 Gabčikovo-Nagymaros Project (Hungary v. Slovakia), Judgment, ICJ Reports 1997, p. 7, para. 54.
21 Pulp Mills on the River Uruguay (Argentina v. Uruguay), Judgment, ICJ Reports 2010, Judges Al-Khasawneh and Simma join Dissenting Opinion, para. 3.
22 Whaling in the Antarctic (Australia v. Japan: New Zealand intervening), Judgment ICJ Reports 2014, p. 226, para. 32.

23 Ibid., para. 33.
24 Ibid.
25 Whaling in the Antarctic (Australia v. Japan: New Zealand intervening), Judgment ICJ Reports 2014, p. 226, Dissenting Opinion of Judge Owada, paras 24 and 25.
26 Mangel (n 18).
27 Certain Activities Carried Out by Nicaragua in the Border Area (Costa Rica v. Nicaragua) and Construction of a Road in Costa Rica along the San Juan River (Nicaragua v. Costa Rica), Judgment, ICJ Reports 2015, p. 665, paras 168–175.
28 See Kanhanga (n 15) at 136.
29 For a recent overview, see Christophe Sobotta, 'The European Union Legal Boundaries for Semi-Natural Habitats Management in Natura 2000 Sites' (2018) 43 *Journal for Nature Conservation* 261.
30 See also Hendrik Schoukens and Kees Bastmeijer, 'Species Protection in the European Union: HowStrict is Strict?' in Charles-Hubert Born et al (eds.) (n 8), 121.
31 Yaffa Epstein et al, 'A Legal-Ecological Understanding of the Favorable Conservation Status for Species in Europe' (2015) 9(2) *Conservation Letters* 81.
32 See for instance Article 11 of Directive 2011/92/EU of the European Parliament and of the Council of 13 December 2011 on the assessment of the effects of certain public and private projects on the environment (2011) OJ L 26/1.
33 Case C-240/09 *Lesoochranárske zoskupenie VLK* EU:C:2011:125, para. 50.
34 Case C-C-115/09 *Trianel* EU:C:2011:289, para. 37.
35 See for instance: Case C-137/14 *Commission v Germany* EU:C:2015:683.
36 Case C-243/15 *Lesoochranárske zoskupenie VLK* EU:C:2016:838, paras 66–70.
37 See more in depth: Jan Darpö, 'Understanding the Nuts and Bolts: Scientific and Technical Knowledge in Environmental Litigation – National Solutions, EU Requirements and Current Challenges', in Lorenzo Squintani et al (eds.), *Managing Facts and Feelings in Environmental Governance* (Edward Elgar 2019) 82.
38 The Aarhus Convention: An Implementation Guide, Second Edition (2014), at 199.
39 See in a similar vein European Commission, Communication from the Commission of 28.4.2017, Commission Notice on Access to Justice in Environmental Cases, C(2017) 2616 final, para. 127 et seq.
40 Case C-72/12 *Gemeinde Altrip* EU:C:2013:712, paras 36–39.
41 European Commission (n 39), at para. 144 et seq.
42 Case C-127/02 *Waddenzee* EU:C:2004:482, para. 59.
43 European Commission (n 39), at para. 145.
44 Aarhus Convention Compliance Committee, Findings and recommendations with regard to communication ACCC/C/2013/90 concerning compliance by the United Kingdom of Great Britain and Northern Ireland. Adopted by the Compliance Committee on 26 July 2021.
45 Case C-237/07 *Janecek* EU:C:2008:447, paras 42–47.
46 Case C-404/13 *ClientEarth* EU:C:2014:2382, paras 55–58.
47 Case C-165/09 *Natuur en Milieu* EU:C:2011:348, paras 100–103.
48 European Commission (n 39), at para. 146–150.
49 For an overview, see Hendrik Schoukens and Hans Woldendorp, 'Site Selection and Designation Under the Habitats and Birds Directives: A Sisyphean Task', in Charles-Hubert Born et al (eds.) (n 8), 31.
50 Case C-355/90 *Commission v Spain* EU:C:1993:331, para. 26.
51 Case C-44/95 *Regina v Secretary of the State for the Environment* EU:C:1996:297, para. 27.
52 While the CJEU's decision in Case C-57/89 *Leybucht* EU:C:1991:89 held that economic considerations could not trump ecological principles when it came to the protection of SPAs for birds, this decision was later effectively overruled by the introduction of the Article 6(4) derogation procedure in the Habitats Directive.
53 Case C-3/96 *Commission v the Netherlands* EU:C:1998:238, paras 68–70.
54 See for instance Case C-235/04 *Commission v Spain* EU:C:2007:386, paras 23–24.
55 Schoukens and Woldendorp (n 49), at 37.

56 Case C-418/04 *Commission v Ireland* EU:C:2007:780, para. 52.
57 Ibid., paras 53–54.
58 Schoukens and Woldendorp (n 49), at 38.
59 Ibid.
60 Volker Mauerhofer, 'Ignorance, Uncertainty and Biodiversity: Decision-Making by the Court of Justice of the European Union', in Christina Voigt (ed.) (n 15), at 153.
61 Case C-334/04 *Commission v Greece* EU:C:2007:628, para. 59.
62 Schoukens and Woldendorp (n 49), at 51.
63 Case C-209/04 *Commission v Austria* EU:C:2006:195, para. 43.
64 Schoukens and Woldendorp (n 49), at 49.
65 Case C-301/12 *Cascina Tre Pini ss* EU:C:2014:214, paras 27–35.
66 Case C-281/16 *Vereniging Hoekschewaards Landschap* EU:C:2017:774, para. 33.
67 Article 2(1) of the Habitats Directive.
68 Epstein et al (n 31). See also Yaffa Epstein, 'Favourable Conservation Status for Species: Examining the Habitats Directive's Key Concept Through a Case Study of the Swedish Wolf' (2016) 28 *Journal of Environmental Law* 221.
69 Epstein et al (n 31), at 82.
70 See for the most ground-breaking decision: Case C-103/00, Commission v Greece ECR I-01147. For an overview of the recent case law developments, see Schoukens and Kees Bastmeijer (n 30) at 121 et seq.
71 Case C-342/05 *Commission v Finland* EU:C:2007:341, para. 29
72 Ibid., paras 33–40.
73 Ibid., para. 38.
74 Case C-383/09 *Commission v France* EU:C:2011:369.
75 For more background, Hendrik Schoukens, 'Going beyond the Status Quo: Towards a Duty for Species Restoration under EU Law?' in Vasilka Sanci and Maša Kovič Dine (eds.), *International Law: Contemporary Concerns and Challenges in 2014* (GV Založba 2014) at 352.
76 Case C-383/09 *Commission v France* EU:C:2011:23, Opinion A.G. Kokott, para. 37.
77 Case C-383/09 *Commission v France* (n 74), para. 24.
78 Mark L. Shaffer, 'Minimum Viable Populations for Species Conservation' (1981) 31 *Bioscience* 131.
79 Lochran W. Traill et al, 'Pragmatic Population Viability Targets in a Rapidly Changing World' (2010) 143 *Biological Conservation* 29
80 I.R. Franklin and R. Frankham, 'How Large Must Populations Be to Retain Evolutionary Potential?' (1998) 1 *Animal Conservation* 69.
81 Ibid.
82 European Commission, 'Reasoned Opinion in Infringement Proceeding 2010/4100' (2015).
83 Ruling of the Swedish Supreme Administrative Court of 30 December 2016, cases 2406-2408-16 & 2628-2630-16, at 22.
84 Case Waddenzee (n 42) paras 47–48.
85 Ibid., para. 54.
86 Case C-241/08 *Commission v France* EU:C:2010:114, para. 53.
87 Case C-258/11 *Sweetman et al* EU:C:2013:220, para. 48.
88 Ibid., para. 45.
89 Case C-164/17 *Grace and Sweetman* EU:C:2018:593, para. 36.
90 Cases C-293/17 and C-294/17 *Coöperatie Mobilisation for the Environment UA* EU:C:2018:882, para. 103.
91 Opinion Advocate General Kokott, Case C-281/16 *Vereniging Hoekschewaards Landschap* EU:C:2017:774, para. 54.
92 Ibid., para. 39.
93 For a more extensive analysis of the precautionary principle, see Peter-Tobias Stoll, 'Of Fear and Prudence: Precaution Through Better Regulation and Innovation', in Lorenzo Squintani et al (n 37) 137.

94 Ibid., at 141.
95 Case Waddenzee (n 42), paras 56–60.
96 See Ralph Frins and Hendrik Schoukens, 'Balancing Wind Energy and Nature Protection: From Policy Conflicts Towards Genuine Sustainable Development', in Lorenzo Squintani et al (eds.), *Sustainable Energy United in Diversity – Challenges and Approaches in Energy Transition in the European Union* (Europa Law Publishing 2014) 84.
97 Case C-43/10 *Nomarchiaki Aftodioikisi Aitoloakarnanias* EU:C:2012:560, para. 115.
98 See for instance: Case C-304/05 *Commission v Italy* EU:C:2007:532, para. 69.
99 See also Advocate General Kokott, Case C-127/02 *Waddenzee* EU:C:2004:60, para. 107.
100 Order, Case C-441/17 R *Commission v Poland* EU:C:2017:877, para. 54.
101 Ibid., para. 55.
102 Ibid., para. 60.
103 Ibid.
104 Ibid., para. 61.
105 Case C-441/17 *Commission v Poland* EU:C:2018:255, para. 173 et seq.
106 Ibid., para. 179.
107 For more background, see Hendrik Schoukens, 'Habitat Restoration Measures as Facilitators for Economic Development Within the Context of the Habitats Directive: Balancing No Net Loss with the Preventive Approach?' (2017) 9(1) *Journal of Environmental Law* 47.
108 See for instance: Michael Curran et al, 'Is There Any Empirical Support for Biodiversity Offset Policy' (2014) 24(4) *Ecological Applications* 617. See also David Moreno-Mateos et al, 'Structural and Functional Loss in Restored Wetlands Ecoystems (2012) 10(1) *PLOS Biol*ogy e1001247.
109 Case C-521/12 *Briels* EU:C:2014:330, para. 26 et seq.
110 Ibid., para. 32.
111 Cases C-387/15 and C-388/15 *Orleans* (2016) EU:C:2016:583, para. 51 et seq.
112 Case Grace and Sweetman (n 90), para. 53.
113 Case C-142/16 *Commission v Germany* EU:C:2017:301, paras 37–45.
114 Case *Coöperatie Mobilisation for the Environment UA* (n 91).
115 For more background, see Hendrik Schoukens, 'Nitrogen Deposition, Habitat Restoration and the Habitats Directive: Moving Beyond the Deadlock with the Dutch Programmatic Nitrogen Approach?' (2017) *Biological Conservation* 484.
116 Case *Coöperatie Mobilisation for the Environment UA* (n 91), para. 96
117 Ibid., para. 101.
118 Ibid., para. 130.
119 See for instance: Cathy J. Stevens, 'How Long Do Ecosystems Take to Recover from Atmospheric Nitrogen Deposition?' (2016) *Biological Conservation* 160.
120 See more elaborately: Esptein (n 9).
121 Adrian Treves et al, 'Longitudinal Analysis of Attitudes towards Wolves' (2013) 27(2) *Conservation Biology* 315.
122 Robert B. Wielgus and Kailey A. Peebles, 'Effects of Wolf Mortality on Livestock Depredations' (2014) 9(12) *PLOS ONE* e113505.
123 Opinion Advocate General Kokott, Case C-342/05 *Commission v Finland* EU:C:2006:752, para. 36.
124 Case *Commission v Finland* (n 71), para. 41 et seq.
125 Case C-674/17 *Tapiola* EU:C:2019:851, para. 67.
126 Ibid., para. 45.
127 Ibid., para. 59.
128 Ibid., paras 48–50.
129 Ibid., para. 66.
130 See Eric Barbier De La Serre and Anne-Lise Sibony, 'Expert Evidence Before the EC Courts' (2008) 45 *Common Market Law Review* 941.

131 Ibid., 950–964.
132 Also Ibid., at 951.
133 Dutch Council of State (May 29, 2019) NL:RVS:2019:1764.
134 See more extensively: Crawford S. Holling (ed.), *Adaptive Environmental Assessment and Management* (John Wiley & Sons 1978).
135 For a more detailed analysis: Hendrik Schoukens, 'The Quest for The Holy Grail and the Dutch Integrated Approach to Nitrogen: How to Align Adaptive Management Strategies with the EU Nature Directives?' (2018) *Journal for European Environmental and Planning Law* 171.
136 On Wikipedia the concept 'citizen science' is defined as follows: scientific research conducted, in whole or in part, by amateur (or non-professional) scientists. See https://en.wikipedia.org/wiki/Citizen_science (accessed 22 September 2022).
137 Margaret Kosmala et al, 'Assessing Data Quality in Citizen Science' (2016) *Frontiers in Ecology and the Environment* https://doi.org/10.1002/fee.1436.
138 Belgian Constitutional Court (April 28, 2016), no. 57/2016.
139 See more extensively: Hendrik Schoukens, 'Towards a Legally Enforceable Duty to Restore Endangered Species under EU nature Conservation Law. On Wild Hamsters, the Rule of Law and Species Extinction', in Jerzy Jendroska and Magdalena Bar (eds.), *Procedural Environmental Rights: Principle X in Theory and Practice* (Intersentia 2018) 287.
140 Administrative Court of Toulouse (2018) case nos. 150.1887 and 102320.
141 See more extensively: Mikael Schultz, 'Scientific Evidence in Swedish Courts: The Use of Technical Judges for Better Integration of Scientific Data in Decision-Making', in Lorenzo Squintani et al (eds.) (n 37) at 118.

10

STRICT PROTECTION OF SPECIES IN THE EU

Controversies and trends

Yaffa Epstein, José Vicente López-Bao and Guillaume Chapron

Introduction

Council Directive 92/43/EEC on the conservation of natural habitats and of wild fauna and flora is often referred to as simply the "Habitats Directive," and indeed, much of the emphasis in its implementation has been on establishing the Natura 2000 network for habitat conservation.[1] Like the Birds Directive though,[2] the Habitats Directive also requires the direct protection of species. This protection in turn entails the indirect protection of species' habitats, whether or not the habitat is located in a protected area.

The Habitats Directive requires EU Member States to take measures to "maintain or restore, at favourable conservation status, natural habitats and species of wild fauna and flora of Community interest."[3] "Species of Community interest" are defined in Article 1(g) of the Directive as those that are endangered, vulnerable, rare, or endemic. Which species may meet this description is not a question that is evaluated by individual Member States however; they are named in annexes to the Directive. Species may be listed in Annex II, "Animal and plant species of Community interest whose conservation requires the designation of special areas of conservation." They may also be listed in Annex IV, "Animal and plant species of Community interest in need of strict protection," or Annex V, "Animal and plant species of Community interest whose taking in the wild and exploitation may be subject to management measures." Most of the species that are listed in Annex II are also listed in Annex IV. Annex IV includes approximately 360 vertebrates, 140 invertebrates, and 700 plant species.[4] The selection of listed species has been a political as well as scientifically informed process, and while many of the species included are "red listed" as threatened or endangered, others are considered to be "of least concern". Insects and other invertebrates

DOI: 10.4324/9780429299100-10

have been argued to be insufficiently represented,[5] as have fungi[6] and marine species.[7]

Which species to include and at what level to protect them has often been the subject of extensive negotiations.[8] Due largely to Member States' opportunities to negotiate when joining the EU, some species that are listed in Annex IV in some regions of the EU are listed in Annex V in others,[9] for example, the European hamster (*Cricetus cricetus*) is listed in Annex IV except in Hungary, where it is listed in Annex V. Species can also be listed in one annex in one part of a country and another annex in another, for example, wolves (*Canis lupus*), are listed in Annex IV in parts of Finland, Greece, and Spain and Annex V in other parts. The Habitats Directive's provisions on species protection, Articles 12–16, pertain to those species listed in Annex IV and Annex V.

Legal controversies around the protection of species have included how to define "favourable conservation status", the system of strict protection, and when exceptions from strict protection may be made. Additionally, litigation has revolved around enforcement questions. Writing in 2015, Schoukens and Bastmeijer noted the strict interpretation of the Habitats Directive's species protection provisions by the CJEU, as well as increasing litigation to enforce the Directive before national courts.[10] These trends have continued, as court decisions such as the *Białowieża Forest* case on the removal of timber in an old-growth forest in Poland,[11] the *Tapiola* case on wolf hunting in Finland,[12] and the *Skydda Skogen* case on logging in Sweden[13] illustrate. In this chapter, we describe the requirements for the protection of species in the Habitats Directive, trends in the interpretation and enforcement of these requirements, and ongoing open questions.

Favourable conservation status for species

The Habitats Directive's key obligation for Member States is to maintain or restore the favourable conservation status of species as well as natural habitats. While conservation status of species and habitats are evaluated separately, they are closely connected. For a natural habitat's conservation status to be considered favourable, the conservation status of its "typical species" must also be considered favourable, amongst other factors.[14] The conservation status of a species is defined as favourable when its habitats are sufficiently large to maintain its populations on a long-term basis. Additionally, population dynamics data must indicate that the species is "maintaining itself on a long-term basis as a viable component of its natural habitats," and its natural range is "neither being reduced nor is likely to be reduced for the foreseeable future."[15] As European Commission guidance documents explain, this definition requires an assessment of the species' population dynamics, range, habitat sufficiency, and prospects for long-term viability.[16]

Member States report the conservation status of natural habitats and species to the European Commission every six years, and the Commission thereafter prepares a report on the state of nature in the EU.[17] An accurate and consistent interpretation of "favourable conservation status" therefore is necessary to assess whether the Directive is being complied with and what its impact is on European nature. It is also important to evaluate conservation status in applying several aspects of the Directive's requirements both with respect to habitat conservation and with respect to species conservation. For example, as will be discussed below, exceptions to the strict protection of species may be made only after an analysis of the species' populations' conservation status.

While the definition of favourable conservation status for species contains many unclear terms,[18] it has been described by the European Commission as "a situation where a … species is doing sufficiently well in terms of quality and quantity and has good prospects of continuing to do so in the future."[19] One significant uncertainty is on what scale conservation status for species should be evaluated when determining favourability. The Habitats Directive aims at ensuring biodiversity "in the European territory of the Member States" through measures to be taken by the individual Member States to achieve and maintain favourable conservation status.[20] It is not explicit in the Directive whether this status must be obtained at the EU, biogeographical, Member State, or local level.[21] This question is significant because it leaves unclear the extent to which each Member State must contribute to a species' favourability at the EU level, and to what extent the Member State can rely on species' populations outside its borders to help meet its obligations.

This question was partially clarified by the CJEU in the *Tapiola* case, at least with respect to identifying the level at which conservation status should be assessed when allowing exceptions from species protection. The answer was that multiple levels must be considered: the Court ruled that when assessing the impact of an exception, it must evaluate the impact on the favourability of the conservation status of the population at both the local level and at the Member State level or, if there are multiple biogeographical regions within the Member State, at the biogeographical level, as well as to the extent possible at the cross-border level.[22] This requirement for multi-level assessment helps ensure that Member States neither shirk their responsibilities by relying on healthy populations in other states nor by focusing only on local populations to the exclusion of promoting healthy populations on a larger scale.[23] Further, individuals of a species in non-EU countries cannot be counted towards the favourability of the conservation status of EU populations of the species; e.g. Finland could not rely on Russian wolf populations when evaluating the conservation status of Finnish wolves, as third countries have no duty of protection towards the species that the EU has designated of Community interest.[24]

The system of strict protection

Member States are required to take measures to establish a system of strict protection for the species listed in Annex IV. For animal species, the contours of this system are described in Article 12 and must include the prohibition of most types of deliberate harm to individuals of the species, including banning deliberate capture, killing, disturbance, and taking of eggs. The CJEU has clarified that deliberate action is not limited to situations where the actor desires to harm the animal in question, but also extends to situations where the actor is aware that the harm could occur.[25] This includes activities that have "manifestly different" purposes than harming species, such as forestry or other economically productive land uses.[26] Therefore, if an environmental impact assessment indicates that an activity may harm individual members of a strictly protected species, carrying out the activity would be considered to be deliberate harm even though the harm is not specifically desired.[27] Additionally, the deterioration and destruction of breeding sites and resting places must be banned whether or not deliberate. This imposes an additional responsibility on Member States to ensure the protection of these areas.[28] These areas must be protected even when not occupied by the species, if there is a "sufficiently high probability" that the species may occupy them in the future.[29] Measures required have included positive measures, such as continuing agricultural practices needed to maintain hamster burrows,[30] and negative measures, such as not removing dead wood that had been colonized by protected beetles.[31] These prohibitions required by Article 12 thus indirectly require the strict protection of species' habitats even outside of Natura 2000 areas, regardless of how plentiful the species may be within Natura 2000 areas.[32]

As noted above, the required system of strict protection includes positive measures as well as prohibitions on harm. Precisely what positive measures are required is less clear than the prohibitions, but must, according to Article 12(4), at least include the establishment of a system of monitoring the incidental capture and killing of these species, as well as the implementation of the additional measures necessary to limit the negative impact of this incidental capture and killing. The CJEU has further interpreted Article 12 to require Member States not only to enact a "comprehensive legislative framework" for species protection but also to implement the concrete measures needed to actually ensure protection.[33] Further, these "coherent and coordinated measures" must in fact prevent harm.[34]

The system of strict protection must be applied regardless of whether species have achieved favourable conservation status. While understanding favourable conservation status is important for evaluating whether the Habitats Directive's goals are met, achieving this status does not remove Member States' Article 12 obligations, as the CJEU clarified in the *Skydda Skogen* case. That case concerned the logging of nearly all trees in an area in which the strictly protected moor frog (*Rana arvalis*) was present, as well as a number of bird species that are protected by the Birds Directive and, under Swedish law, were given the same level of protection as species that are strictly protected by the Habitats Directive.[35] Interpreting Article 12 of the Habitats Directive, the CJEU confirmed that the prohibitions included in the required system of strict protection continue to apply to species that have favourable conservation status.[36] Further, the deterioration or destruction of breeding sites and resting places of strictly protected species must be prevented regardless of how many specimens of the species might be affected, and regardless of whether the impact on the site would be likely to cause the conservation status of the species to deteriorate.[37]

Exceptions from species protection

Like many environmental protection laws, the Habitats Directive allows exemptions from its restrictions in certain limited circumstances. Mirroring the Birds Directive's Article 9(1), Article 16 of the Habitats Directive delimits when Member States are allowed to "derogate" from its provisions on strict protection. There are three conditions, all of which must be fulfilled, in order to derogate from the other articles relating to species protection, Articles 12–15. These three conditions are: 1. There must be no satisfactory alternative to derogation, 2. Derogation must not be detrimental to the maintenance of the populations of the species at a favourable conservation status in their natural range, and 3. The derogation must be for one of the following five reasons:

a in the interest of protecting wild fauna and flora and conserving natural habitats;
b to prevent serious damage, in particular to crops, livestock, forests, fisheries and water, and other types of property;
c in the interests of public health and public safety, or for other imperative reasons of overriding public interest, including those of a social or economic nature and beneficial consequences of primary importance for the environment;
d for the purpose of research and education, of repopulating and re-introducing these species, and for the breedings operations necessary for these purposes, including the artificial propagation of plants;
e to allow, under strictly supervised conditions, on a selective basis and to a limited extent, the taking or keeping of certain specimens of the species listed in Annex IV in limited numbers specified by the competent national authorities.

It is necessary to identify a reason for derogating from amongst these five before analysing the other two conditions. Without identifying a reason, it is not possible to analyse whether satisfactory alternatives exist. Further, it would be impossible to know the course of action needed to achieve the purpose, and therefore to analyse the impact of the necessary action on the conservation status of the species' population.

When derogations may be made to allow for the capture or killing of individual animals of strictly protected species has been particularly controversial. The CJEU has interpreted Article 16 with respect to this issue in two cases involving wolf hunting in Finland. The first of these, the 2007 *Commission v Finland* case,[38] involved the hunting of wolves under derogation ground (b), the prevention of serious damage to property, while the 2019 *Tapiola* case[39] involved derogation ground (e), which does not in fact describe a purpose but rather an additional set of limitations.

In *Commission v Finland*, the CJEU noted the standards by which it would evaluate exceptions to strict protection, which apply to each of the five reasons set out above. Citing its earlier *WWF Italia and Others* decision interpreting the corresponding provision of the Birds Directive,[40] the CJEU stated that because Article 16 of the Habitats Directive pertained to exceptions from a coherent system of species protection, this article must be interpreted strictly and Member States had the burden of proving that all conditions for derogating were met. Member States were required to provide a "clear and sufficient" statement as to why the derogation was necessary and how the requirements for granting it were fulfilled.[41] Noting that purpose (b) allowed derogation to *prevent* serious damage, the CJEU ruled that a derogation could be allowed for this reason even if serious damage had not yet occurred.[42] However it was necessary for Member States to demonstrate the connection between the derogation and the prevention of serious damage.[43]

The level of proof required was further developed in the *Tapiola* case. After losing the 2007 case because it had not demonstrated the connection between hunting and the prevention of serious damage, Finland began instead granting hunting permits under derogation ground (e). As noted, this derogation ground does not state a particular purpose, but rather imposes a stricter set of conditions when granting derogations for other reasons. Finland claimed its purpose for derogating was to prevent poaching, as well as to prevent serious damage to dogs and livestock, and to alleviate the fear of wolves amongst local inhabitants. Two of these hunting permits were appealed to the Finnish administrative court, and then to the Finnish Supreme Administrative Court, which in turn referred several questions relating to the legality of Finland's derogation to the CJEU for a preliminary ruling.

The CJEU again confirmed that Article 16 must be interpreted restrictively, noting that it defined "in a precise and exhaustive manner" the circumstances under which Member States may derogate from the provisions on species protection.[44] The CJEU went on to analyse aspects of all three conditions for granting derogations. Going further than its earlier ruling, the Court emphasized the high level of scientific evidence required to demonstrate that each condition was met.

Before this case, it had been unclear whether Member States were limited in any way as to the purposes that may be relied upon under derogation ground (e). The CJEU has now clarified several restrictions on utilizing this derogation ground. First, it can only be used when the purposes set out under grounds (a)–(d) are irrelevant.[45] Otherwise, the other grounds for derogation would be unnecessary.[46] Further, derogations made under any of the five derogation grounds may not have results that conflict with the Habitats Directive's conservation objectives.[47]

Importantly, the court discussed the level of proof required to demonstrate that the derogation was appropriate to its intended purpose. The intended purpose of preventing poaching was found to be permissible. While no evidence that legal hunting could prevent poaching was supplied, Finland argued that it was testing the proposition. Given the uncertainty, the CJEU emphasized the need for Member States to supply "rigorous scientific evidence" that such a derogation would have a net positive impact on the species concerned.[48] When there is uncertainty about the effectiveness of a management measure, a high level of scientific evidence must be presented to demonstrate that derogation could lead to favourable outcomes for the species.

The CJEU then went on to analyse the condition that no satisfactory alternative exists. Member States must choose the least harmful means for achieving their objectives. Expanding on its earlier requirement for a clear and sufficient statement of reasons, the CJEU indicated that Member States must also take into account the best available scientific and technical reports to support their conclusion that derogation was the only satisfactory way to achieve the stated goal.[49] Again, the Court emphasized a requirement for Member States to show convincing scientific evidence when claiming a need to make exceptions from strict protection.

Enforcement

Schoukens and Bastmeijer have noted both the important role of the European Commission in bringing proceedings for failure to fulfil obligations in environmental cases under Article 258 of the TFEU, as well as the increasing importance of non-governmental actors in enforcing EU environmental law in the national courts.[50] Both of these trends have continued: the role of the national courts in ensuring compliance with the Habitats Directive and EU environmental law more generally has continued to increase, and while the European Commission has reduced its number of direct legal actions, it continues to work to promote enforcement through dialogue, capacity building, guidance documents, and strategically selected cases.[51] The *Białowieża Forest* case and the aforementioned *Tapiola* case illustrate the successes and differences between these two routes to enforcement.

The *Białowieża Forest* case was brought by the European Commission after complaints from individuals and environmental organisations.[52] In that case, Poland permitted timber removal in an old growth forest that is a Natura 2000 site as well as a breeding site and resting place for several beetles that are listed in Annex IV of the Habitats Directive. Poland claimed that the continued removal of timber was a conservation measure meant to control the spread of the spruce bark beetle (*Ips typographus)*, a non-protected species that can be deadly to spruce trees, especially those that have been weakened by drought.[53] However, the European Commission and environmental advocates argued, the supposed conservation measures were in fact designed to produce profits from timber removal rather than to promote conservation, and would have a deleterious rather than beneficial impact on the forest and its habitats. The CJEU ruled that Poland had violated several provisions of the Habitats Directive as well as the Birds Directive, including Article 12(1) (a) and (d) of the Habitats Directive for failing to guarantee the strict protection of Annex IV beetles, and for failing to ensure the protection of certain birds under similar provisions of the Birds Directive.

The *Tapiola* case was initiated by a regional Finnish NGO that was formed in 2014 for the express purpose of challenging Finland's wolf hunting laws and policies. Most cases involving EU law are resolved in national courts, and only a small fraction of these cases reach the CJEU through references for a preliminary ruling. Still, the CJEU's caseload now consists of far more cases initiated in the national courts rather than direct actions brought by the European Commission:[54] In 2021, the CJEU ruled in 547 cases referred by national courts and only 30 direct actions.[55] Perhaps because environmental cases often involve public rather than private interests, the European Commission has remained relatively active in bringing cases in this area, but of the new cases lodged with the CJEU in 2021, still there were significantly more references for a preliminary ruling as there were direct actions in environmental cases, 12 versus 8.[56] The *Tapiola* case may therefore be considered part of the trend towards the increasing importance of national courts in resolving matters of EU environmental law.

Like other Nordic countries, Finland does not have a strong tradition of public interest litigation;[57] however, the possibilities for organizations to bring public interest environmental litigation have increased to comply with EU requirements.[58] In 2015, Finland granted permits to kill 24 wolves, and Tapiola appealed them all to the Finnish administrative courts on the grounds that killing wolves violated Articles 12 and 16 of the Habitats Directive. While the Finnish Hunting Act allows local and regional associations whose purpose is to promote nature or environmental protection to appeal decisions to hunt species that are protected by the Habitats Directive,[59] Tapiola was not successful. Most of the appeals were rejected on the grounds that Tapiola's registered address was too far away from the areas where the permits had been granted for it to be considered a local organization.

The following year, Tapiola reorganized into six regional organizations to appeal permits granted in 2016 allowing for the killing of 46 wolves. This time, most of the administrative courts granted these regional organizations standing to bring their claims, but standing was denied by the Eastern Finland Administrative Court for the two appeals filed in that court. Tapiola appealed these two rejections to the Finnish Supreme Administrative Court. It also requested that the Supreme Administrative Court seek a preliminary ruling on the issues of EU law arising. Referring to decisions of the CJEU in the *Slovak Brown Bear* cases, which required national courts to allow broad standing to challenge violations of EU environmental law,[60] the Supreme Administrative Court ruled that Tapiola did have standing in these cases, and referred the questions of EU law to the CJEU for a preliminary ruling.[61]

Both the *Białowieża Forest* and *Tapiola* cases demonstrate the importance of non-governmental actors, as well as of the European Commission in bringing about compliance with EU law. In the *Tapiola* case, an NGO action to enforce the Habitats Directive applied the broad standing now available to bring litigation to enforce EU environmental law at the national level. Though it did not bring the case, the European Commission also played an important role, supporting Tapiola's arguments by submitting both written and oral observations to the CJEU as part of the preliminary reference proceedings. The resulting CJEU decision resulted in a strict interpretation of aspects of Article 16 of the Habitats Directive that is applicable throughout the EU.

The increasing possibilities for non-governmental actors to bring claims in environmental cases at the national level is one reason for the increase in references for preliminary rulings as compared to direct actions brought by the European Commission. The Commission's stated policy of focusing its direct actions on situations where national legal systems make it difficult for non-governmental actors to bring about the enforcement of EU law is another.[62] The *Białowieża Forest* case, in which Poland was seen as wilfully flouting the rule of law, was one such situation.[63] Here, although the case was a direct action, non-governmental actors also were instrumental in bringing about compliance with the Habitats Directive by providing information about the ongoing violations to the European Commission. The European Commission itself does not have the apparatus to monitor Member State compliance with EU environmental law, instead relying on members of the public to make complaints about non-compliance.[64] In the *Białowieża Forest* case, complaints by the public as well as whistleblowers within the Polish administration made it possible for the European Commission to prosecute its case.[65] While the *Tapiola* and *Białowieża Forest* cases took different routes to the CJEU, both the European Commission and non-governmental actors contributed to bringing about the legal enforcement of the Habitats Directive.

It is useful though to note the different types of result possible in a reference for a preliminary ruling versus a direct action. In the *Tapiola* case, a preliminary reference, the CJEU promulgated a strict interpretation of the Habitats Directive that is applicable throughout the EU, but it could not rule on whether Finland violated the law. It is for the national courts to examine the facts and determine whether the law has been violated, applying the CJEU's interpretation of the questions of EU law referred. This requires national courts that are willing and able to enforce EU law, which cannot always be taken for granted. In the *Białowieża Forest* case, a direct action, the European Commission investigated and presented evidence of violations of the Habitats Directive. The CJEU examined this evidence and determined that Poland was in violation of the Habitats Directive. While the case was ongoing, the CJEU also took the unusual step, at the request of the European Commission, of granting interim measures ordering Poland to refrain from removing timber while the case was pending, and threatened steep monetary penalties when Poland continued to carry out logging.[66] While the European Commission currently brings only a relatively small number of direct actions, its ability to do so remains an important "stick" to encourage compliance.

Continuing issues

The decisions in the *Białowieża Forest* and *Tapiola* cases have clarified several important questions on the system of strict protection for species required by the Habitats Directive. They raise others. This section considers some issues remaining after these cases.

Precautionary principle

The decisions in both the *Białowieża Forest* case and the *Tapiola* case emphasized the role of the precautionary principle in EU environmental law. When granting interim measures in the former case, the CJEU stated that the precautionary principle "is one of the foundations of the high level of protection aimed at by EU policy on the environment" that must be applied whenever the facts are unclear.[67] As noted by Krämer, this is a new and strong statement of how far the precautionary principle should be applied in interpreting EU environmental law.[68]

The need for scientific evidence to apply the precautionary principle was also emphasized in the *Tapiola* case when considering the impact on the conservation status of species' populations. The court ruled that derogations could not be granted if, after an analysis of the best available science, there was significant doubt whether derogation would be detrimental to maintaining or restoring species' populations at favourable conservation status.[69] In its holding, the CJEU appeared to go even further, stating that, at least when the objective of the derogation was to combat poaching, it must be "guaranteed" that derogation would not be detrimental to the species' populations, a very strong statement of the level of precaution to be applied.

The precautionary principle has now been specifically applied to several aspects of the Habitats Directive, and the CJEU has confirmed its importance to the interpretation of EU nature protection law more broadly.[70] The strong statements of the role of this principle may have wide-reaching consequences. For instance, application of the precautionary principle to the management of Annex V species may require stricter oversight of their hunting and other management measures.

Reasons for Article 16(1)(e) derogation

In the *Tapiola* case, the CJEU indicated that fighting poaching could in principle justify derogation from the restrictive provisions of Article 12 of the Habitats Directive, but that it would be very difficult for this to be the case in part because there are likely more satisfactory means to combat poaching. It is an open question whether there are reasons for hunting for which no satisfactory alternative exists. An example might be hunting as a culturally valuable activity in itself, or hunting to preserve other culturally valuable activities such as reindeer husbandry and pastoralism. While it is difficult to conceive of a situation in which the hunting of strictly protected species would be found to be permissible, Member States and hunting organizations will likely continue to seek justifications for hunting that may satisfy the Court.

Amending the Habitats Directive's annexes

A strict interpretation of strict protection has been confirmed by the CJEU. This may lead to future attempts to remove species from strict protection or to find other avenues for "flexible" management. According to Article 19 of the Habitats Directive, the annexes to the Directive may be amended by a qualified majority vote of the Council of the European Union, except for Annex IV, which requires unanimity. This makes it very difficult to remove species from strict protection, even if they are flourishing and arguably no longer fall within the definition of "species of Community interest."

Some Member States have argued for greater flexibility in managing species' populations that have reached favourable conservation status, either by interpreting Articles 12 and 16 less strictly for Annex IV populations that have recovered, or by amending the annexes to move species from Annex IV to Annex V. Wolf populations, in particular, have increased in several Member States since their protection under the Habitats Directive, which was of course the intent of strictly protecting them. Not all Member States with recovering populations are pleased about it, with some proposing to remove wolves from strict protection.[71] But despite some political pressure, the trend seems to continue towards rigorous interpretation and application of the Habitats Directive's provisions on species protection by the Court of Justice and the European Commission.

Notes

1 Hendrik Schoukens and Kees Bastmeijer, "Species Protection in the European Union: How Strict is Strict?" in C.H. Born, A. Cliquet, H. Schoukens, D. Misonne and G. Van Hoorick (eds.), *The Habitats Directive in its EU Environmental Law Context: European Nature's Best Hope?* (Routledge, 2015), 122.
2 Directive 2009/147/EC of the European Paliament and of the Council on the Conservation of Wild Birds.
3 Council Directive 92/43/EEC of 21 May 1992 on the conservation of natural habitats and of wild fauna and flora (OJ L 206, 22.7.1992, p. 7), Article 2.
4 Yaffa Epstein et al, "When Is It Legal to Hunt Strictly Protected Species in the European Union?" (2019) 1 *Conservation Science and Practice* e18, 2.
5 Pedro Cardoso, "Habitats Directive Species Lists: Urgent Need of Revision" (2012) 5 *Insect Conservation and Diversity* 169.
6 Rob Amos, "Assessing the Impact of the Habitats Directive: A Case Study of Europe's Plants" (2021) 33 *Journal of Environmental Law* 365, 370.
7 Douglas Evans et al, "Adapting Environmental Conservation Legislation for an Enlarged European Union: Experience from the Habitats Directive" (2013) 40(2) *Environmental Conservation* 97, 103.
8 Ibid.
9 Ibid.
10 Schoukens and Bastmeijer, n 1.

11 Case C-441/17 *Commission v Poland,* EU:C:2018:255.
12 Case C-674/17 *Luonnonsuojeluyhdistys Tapiola Pohjois-Savo — Kainuu ry,* EU:C:2019:851 (Tapiola case).
13 Joined cases C-473/19 and C-474/19 *Föreningen Skydda Skogen and Others v Länsstyrelsen i Västra Götalands län and Others,* EU:C:2021:166.
14 Habitats Directive, Article 1(e).
15 Ibid.
16 DG Environment, "Reporting under Article 17 of the Habitats Directive: Explanatory Notes and Guidelines for the Period 2013–2018" (2017), 7–8, available at https://cdr.eionet.europa.eu/help/habitats_art17.
17 E.g., European Commission, "The State of Nature in the European Union: Report on the Status and Trends in 2013 – 2018 of Species and Habitat Types Protected by the Birds and Habitats Directives," COM/2020/635 final (15 October 2020).
18 Yaffa Epstein, José Vicente López-Bao and Guillaume Chapron, "A Legal-Ecological Understanding of Favorable Conservation Status for Species in Europe" (2016) 9 *Conservation Letters* 81.
19 European Commission,"Note to the Habitats Committee" (DocHab-04-03/03_rev.3) (2005) 4, available at https://circabc.europa.eu/sd/a/83bc5fcd-7118-4912-a4dc-fce4c00e2512/DocHab-04-03-03%20rev3.doc.
20 Habitats Directive, Article 2.
21 Arie Trouwborst, Luigi Boitani and John DC Linnell, "Interpreting 'Favourable Conservation Status" for Large Carnivores in Europe: How Many Are Needed and How Many Are Wanted?'(2017) 26 *Biodiversity and Conservation* 37.
22 *Tapiola* case, n 12, para. 61.
23 Yaffa Epstein and Guillaume Chapron, "The Hunting of Strictly Protected Species: The Tapiola Case and the Limits of Derogation Under Article 16 of the Habitats Directive" (2018) 27 *European Energy and Environmental Law Review* 78, 86. The requirement for multi-level assessment of conservation status should also apply to the Environmental Libility Directive (2004/35/EC), which uses the same term.
24 Ibid., para. 60.
25 Case C-221/04 *Commission v Spain,* EU:C:2006:329, para. 71.
26 *Skydda Skogen,* supra n 13.
27 Minna Pappila, 'Summer Loggings and Bird Protection: On Regulation and Derogations' (2019) 28(2) *Review of European, Comparative & International Environmental Law* 210, 215.
28 Schoukens and Bastmeijer, n 1, 139.
29 Case C-477/19 *IE v. Magistrat der Stadt Wien,* EU:C:2020:517.
30 Case C-383/09 *Commission v France,* EU:C:2011:369.
31 *Commission v Poland*, supra n 11, paras 233–238.
32 Ibid., 237.
33 Ibid., 231.
34 Ibid.
35 *Skydda Skogen,* supra n 13, paras 27 and 46.
36 Ibid., ruling 2.
37 Ibid., ruling 3.
38 C-342/05 *Commission v Finland,* EU:C:2007:341.
39 Tapiola case, supra n 12.
40 Case C-60/05 *WWF Italia and Others v Regione Lombardia,* EU:C:2006:378, para. 34.
41 *Commission v Finland*, supra n 38, para. 25.
42 Ibid., para. 40.
43 Ibid., para. 47.
44 *Tapiola* case, n 12, para. 30.
45 *Tapiola* case, n 12, para. 37.
46 Ibid., para. 36.

47 Ibid., para. 38.
48 Ibid., para. 45.
49 Ibid., para. 51.
50 Schoukens and Bastmeijer, n 1, 128.
51 European Commission, "Communication from the Commission: EU law: Better results through better application (2017/C 18/02)." See also the chapter in the present book by Liam Cashman on the European Commission's environmental compliance assurance programme.
52 Ludwig Krämer, "Injunctive Relief in Environmental Matters: Court of Justice of the EU, Cases C-441/17R Commission v. Poland, ECLI: EU: C: 2017: 877 and C-441/17 Commission v. Poland, ECLI: EU: C: 2018: 255" (2018) 15 *Journal for European Environmental & Planning Law* 259, 266.
53 Erik Stokstad, "Beetles Are Ravaging Europe's Oldest Forest. Is Logging the Answer?" Science (5 December 2017).
54 Andreas Hofmann, "Left to Interest Groups? On the Prospects for Enforcing Environmental Law in the European Union" (2019) 28 *Environmental Politics* 342.
55 Court of Justice of the European Union, *Annual Report 2021 Judicial Activity* (2022) 236.
56 Ibid., at 233.
57 Helle Tegner Anker and Annika Nilsson, "The Role of Courts in Environmental Law-Nordic Perspectives" (2010) 3 *Journal of Court Innovation* 111, 117.
58 Lisa Vanhala, "Is Legal Mobilization for the Birds? Legal Opportunity Structures and Environmental Nongovernmental Organizations in the United Kingdom, France, Finland, and Italy" (2018) 51 *Comparative Political Studies* 380, 394.
59 Finnish Hunting Act (615/1993, as amended), Section 90.
60 Case C-243/15 *Lesoochranárske zoskupenie VLK v Obvodný úrad Trenčín,* EU:C:2016:838; Case C-240/09 *Lesoochranárske zoskupenie VLK v Ministerstvo životného prostredia Slovenskej republiky,* EU:C:2011:125.
61 Supreme Administrative Court of Finland, Request for a preliminary ruling from the Court of Justice of the European Union, 28 November 2017, KHO:2017:182.
62 European Commission, 'Communication from the Commission: EU Law: Better Results through Better Application' (2017/C 18/02).
63 Editorial Comment, "About Brexit Negotiations and Enforcement Action Against Poland: The EU's Own Song of Ice and Fire" (2017) 54 *Common Market Law Review* 1309.
64 Krämer, n 52, 265.
65 Ibid.
66 Ibid.
67 Case C-441/17 R *European Commission v Poland,* EU:C:2017:877, para. 42.
68 Krämer, n 52, 264.
69 Ibid., para. 66.
70 Christoph Sobotta, "Recent Applications of the Precautionary Principle in the Jurisprudence of the CJEU–A New Yardstick in EU Environmental Decision Making?" (2021) 21 *ERA Forum* 723.
71 Arie Trouwborst, 'The EU Habitats Directive and Wolf Conservation and Management on the Iberian Peninsula: A Legal Perspective" (2014) 26 *Galemys: Spanish Journal of Mammalogy* 15.

11

CONSERVATION OF AREAS OUTSIDE THE NATURA 2000 NETWORK

An historical perspective with an eye to the future

Andrew L. R. Jackson

Introduction

The Natura 2000 network of protected areas – often cited as the centrepiece of EU biodiversity policy[1] – is undoubtedly one of the most consequential achievements of EU environmental policy. The network, created pursuant to the EU's Habitats Directive[2] and Birds Directive,[3] now comprises >27,000 sites, covering about 18% of the EU's terrestrial area and about 9% of its marine area.[4] This means, of course, that the majority of the EU's terrestrial and marine areas fall *outside* Natura 2000. There is little doubt that the Habitats and Birds Directives are among the most powerful and far-reaching the EU has ever passed in the environmental sphere, impacting as they do on day-to-day land use planning across Europe. That said, and as this chapter will show, an important "network" element of the Natura 2000 network is much weaker than the original proposers of Natura 2000 had intended: namely, the protection of corridors and stepping stones between sites, and more generally the protection of areas outside Natura 2000 for the benefit of the network and for EU biodiversity more broadly.[5]

Much conservation effort in Europe has focused over the past decades on establishing the Natura 2000 network and thereafter seeking to ensure appropriate management of Natura 2000 sites. While research has revealed that the network is delivering conservation benefits[6] notwithstanding a poor picture for biodiversity overall in the EU,[7] international headline news has been made recently by reports of dramatic declines in, for example, flying insect abundance (a greater than 75% decline in total insect biomass within protected areas) in Germany between 1989 and 2016,[8] and a decline in farmland bird abundance of at least one-third in France between 2001 and 2018.[9] In both cases, agricultural intensification was cited as a plausible cause, and this resonates with the European Environment Agency's State of Nature in the EU report, which concluded

DOI: 10.4324/9780429299100-11

in 2020 that "agriculture is the most urgent pressure on habitats and species," continuing that

> many terrestrial habitats are severely impacted by agriculture, especially grasslands and freshwater habitats, heath and scrub, and bogs, mires and fens. This is also the case for most of the species groups, including reptiles, molluscs, amphibians, arthropods, vascular plants and breeding birds.[10]

The question of conservation *outside* EU protected areas is therefore a matter of ever-increasing conservation priority, as a matter both of EU policy and law. This is reflected in the EU Biodiversity Strategy for 2030,[11] and in the proposed EU Nature Restoration Law,[12] to which this chapter returns below.

Reforms of (in principle) complementary EU policies have repeatedly failed to deliver in practice for biodiversity in Europe. As the European Commission commented in 2015 regarding financial investment in nature, "the opportunities offered by, for example, the Common Agricultural Policy [CAP], the Common Fisheries Policy [CFP] and the EU Regional Policy have not been fully realised."[13] Previous iterations of the CAP have been criticised for having "failed to reverse biodiversity loss or markedly reduce the environmental footprint of European agriculture," with critics proposing "that this is due to low requirements and broad exemptions in the compulsory instruments, unambitious design of voluntary schemes, overpayment for ineffective environmental measures, and imbalanced investment in the environment compared to other objectives."[14] As the European Court of Auditors noted in a 2020 report,

> the effect of CAP direct payments – 70% of EU agriculture spending – on farmland biodiversity is limited. Some direct payment requirements, notably greening, and cross-compliance, have potential to improve biodiversity, but the Commission and Member States have favoured low-impact options. The EU's rural development instruments have greater potential than direct payments for maintaining and enhancing biodiversity. However, Member States relatively seldom use high impact rural development measures such as result-based and 'dark green' schemes.[15]

While the CAP reform agreed for the 2023–2027 period was undoubtedly disappointing from an environmental perspective, a group of more than 300 experts has produced recommendations regarding how best to implement its "Green Architecture" for biodiversity purposes.[16] For its part, the CFP is currently the subject of a request for a preliminary ruling from the CJEU relating to the Member States' failure to end overfishing by 2020, which may – depending on the outcome of the case – have major implications for the future direction of the fishing industry in Europe.[17]

At the level of EU biodiversity policy, much emphasis has been placed on so-called green infrastructure as a future area of policy development, both to increase the effectiveness of the Natura 2000 network itself and to conserve areas outside Natura 2000. In this regard, the EU's Biodiversity Strategy to 2020 included as Target 2, "By 2020, ecosystems and their services [will be] maintained and enhanced by establishing green infrastructure and restoring at least 15% of degraded ecosystems."[18] The Commission followed up with a green infrastructure strategy in 2013, which amongst other things noted that the development of

> trans-European priority axes for [green infrastructure] in Europe, TEN-G (based on trans-European networks in grey infrastructure sectors), would have significant benefits for securing the resilience and vitality of some of Europe's most iconic ecosystems, with consequential social and economic benefits.[19]

In reviewing progress regarding the Biodiversity Strategy in 2015, the Council of the EU called on the European Commission "to put forward by 2017 a proposal for a trans-European network for green infrastructure in Europe – TEN-G,"[20] equivalent to existing major networks for energy (TEN-E), telecommunications (eTEN) and transport (TEN-T). While this has not yet happened, preparatory research was undertaken for the Commission,[21] and the Council in its conclusions regarding the Commission's 2017 Action Plan for nature, people and the economy reiterated its call on the Commission to put forward a proposal for a trans-European network for green infrastructure.[22] Most recently, the EU Biodiversity Strategy for 2030 called for the building of a "truly coherent Trans-European Nature Network," integrating ecological corridors with the site protection network.[23] In pursuit of this, a consortium of scientists and NGOs from 15 countries will be working over the next several years to "design and develop a blueprint for a truly coherent Trans-European Nature Network (TEN-N)," as part of the EU-funded project, NaturaConnect.[24]

In the absence of a major new policy intervention such as a TEN-N or the recently announced EU Nature Restoration Law, some attention has focused over the past years on examining existing EU laws with a view to their potential use to secure better conservation outcomes for areas outside Natura 2000. In particular, Verschuuren[25] and Van Hoorick[26] have separately explored provisions of the Habitats and Birds Directives as well as other instruments of EU environmental law in this regard. Both authors propose various ways in which EU law could potentially be used more effectively to conserve biodiversity outside Natura 2000 areas.

The present chapter seeks to contribute to discussions in this area by offering an historical perspective on the development of the Habitats and Birds Directives insofar as areas outside Natura 2000 are concerned. By understanding how the relevant provisions of these Directives came to be agreed, and by understanding what was discounted or excluded during negotiations, as well as the positions adopted by different parties, we may perhaps arrive at a clearer understanding of the arguments that have in the past framed and constrained the development of legal protections for areas outside the Natura 2000 network. Similar issues may or may not arise in future discussions regarding the conservation of such areas – for example, during the forthcoming negotiations of the proposed EU Nature Restoration Law – but in understanding the policy dynamics that were at play and the discussions that took place during earlier periods in EU environmental policy, today's policy makers and stakeholders may perhaps be better positioned to secure a positive outcome for biodiversity when the next occasion arises.

The Birds Directive

The publication in 1962 of Rachel Carson's now classic book *Silent Spring* alerted the public at large to serious threats to wild bird populations.[27] Widespread hunting of migratory birds in France, Italy and Belgium, amongst other places, then came to be an issue of major concern in Europe in the late 1960s and 1970s and led many to campaign for EU legislation on bird conservation.[28] However, by the time the Birds Directive proposal emerged in December 1976 – having been adopted by the Commission "without great internal difficulties" according to the Commission's senior negotiator, Claus Stuffmann[29] – the policy aim had shifted in an important respect, with habitat protection firmly rooted in the proposal,[30] due in no small part to the work of then-Commission lawyer and ornithologist John Temple Lang.[31]

"It is clear," the Explanatory Memorandum to the Commission's Birds Directive proposal records,

> that the situation of some 400 species of birds observed regularly in the territory of the Member States is giving rise to serious problems for several reasons...[S]ome sixty species are currently in danger of extinction, whilst not even one third of the species show normal reproduction rates.[32]

As was the case in Stanley Cramp's short scientific report that had informed the Commission's proposal,[33] the first reason cited in the Explanatory Memorandum for this state of affairs was habitat change and habitat loss: "Wildlife cannot be preserved without adequate space for the survival and reproduction of the species."[34]

Notwithstanding the inclusion of habitat protection in the Commission's proposal, the Member States' attention during negotiations remained predominantly focused on hunting control measures rather than the protection of habitats.[35] The negotiations were by no means straightforward, with various thorny issues emerging over the two plus years it took to agree the instrument: e.g. Denmark was reluctant to allow the EU to regulate nature conservation at all; a dispute arose over the legal basis for the proposal; and Member States with strong hunting lobbies (Belgium, France, Italy) remained very sensitive to hunting regulation.[36]

Two provisions of the Birds Directive are of particular interest in the context of protecting areas outside protected areas,[37] and these are discussed in turn below.

Article 3 of the Birds Directive

The Member States discussed the Commission's directive proposal for the first time at a Working Party on the Environment meeting on 11 March 1977.[42] At this time there were just nine Member States: Belgium, Denmark, France, Germany,[43] Ireland, Italy, Netherlands, Luxembourg and the UK. At the negative end of the spectrum, Belgium, Denmark, France and Ireland gave a "cautious reception" to Article 3 as proposed. France and Ireland were not in favour of including "restoring" within Article 3 "because of the different implications such an obligation could have." Belgium wanted "all [species]" deleted "as it seemed unrealistic in view of the geographical and territorial position of Belgium." The UK wanted Article 3 to be qualified in the same way as Article 2, namely by reference to maintaining habitats "at a level compatible with ecological, economic, recreational and scientific requirements,"[44] and this suggestion was effectively adopted in the final text of the Directive (see Table 11.1).

At the more positive end of the spectrum, the German, Netherlands and Italian delegates "could agree in principle to the ideas contained in Article 3." The Netherlands and Italy argued that the obligations in Article 3 "were simply for general guidance," in contrast to other provisions that "contained precise and

TABLE 11.1 Text of Article 3 in the Commission's Birds Directive proposal (December 1976)[38] versus the text of Article 3 in the adopted version of the Birds Directive (April 1979)[39]

Text of Commission's Birds Directive proposal (December 1976)	*Text of adopted version of Birds Directive (April 1979) following negotiations*
Article 3 The Member States shall take the requisite measures to maintain or restore a sufficient diversity and area of habitats for all species of wild bird.	Article 3 1 In the light of the requirements referred to in Article 2,[40] Member States shall take the requisite measures to preserve, maintain or re-establish a sufficient diversity and area of habitats for all the species of birds referred to in Article 1.[41] 2 The preservation, maintenance and re-establishment of biotopes and habitats shall include primarily the following measures: a creation of protected areas; b upkeep and management in accordance with the ecological needs of habitats inside and outside the protected zones; c re-establishment of destroyed biotopes; d creation of biotopes.

detailed obligations." Germany made the most positive contribution, requesting that the general obligation as proposed in Article 3 be clarified in a second paragraph, "with a view to listing the practical means and measures necessary for such action, namely the establishment of protected areas, the restoration of biotopes and habitats which have been destroyed, etc." Germany later fleshed out this suggestion in more detail,[45] and its ideas were ultimately adopted with only minor changes, resulting in the second paragraph of the provision as finally adopted (see Table 11.1).

Article 4(4), second sentence

On 20 November 1977, the Council's Working Party on the Environment provided a summary of outstanding issues regarding the Birds Directive proposal to the Committee of Permanent Representatives.[48] The draft directive attached to this summary reveals that the habitat protection provisions in what was by now Article 4(4) had been strengthened (with no record apparent of the discussions which led to this[49]) by the addition of a second sentence to the end of Article 4(4) reading: "Outside the special protected areas, Member States shall also make every effort to prevent pollution or deterioration of habitats."[50] This was amended in the final adopted version of the directive to read, "Outside these protection areas, Member States shall also strive to avoid pollution or deterioration" (see Table 11.2).[51]

Thus, already in the Birds Directive, the EU legislature was clearly thinking beyond protected areas to the wider landscape, an issue that would assume much greater significance in the context of the Habitats Directive negotiations. As Meyer comments, "[t]he Birds Directive was originally motivated by Northern

TABLE 11.2 Text of Article 4(3) in the Commission's Birds Directive proposal (December 1976)[46] versus the text of Article 4(4) in the adopted version of the Birds Directive (April 1979)[47]

Text of Commission's Birds Directive proposal (December 1976)	*Text of adopted version of Birds Directive (April 1979) following negotiations*
Article 4 3 The Member States shall take appropriate measures to prevent, in the special protected areas referred to in paragraphs 1 and 2 of this Article, the pollution and deterioration of habitats and the disturbance of birds, should these factors have a significant effect having regard to the objectives of this Article.	Article 4 4 In respect of the protection areas referred to in paragraphs 1 and 2, Member States shall take appropriate steps to avoid pollution or deterioration of habitats or any disturbances affecting the birds, in so far as these would be significant having regard to the objectives of this Article. Outside these protection areas, Member States shall also strive to avoid pollution or deterioration of habitats.

European resentment against the hunting of songbirds in Southern Europe," but ended up introducing, in addition to its hunting control measures,

> remarkably strict rules on habitats protection valid in all member states. It provided an important precedent for EC action in the area of nature protection and paved the way for the Habitats directive of 1992 and the Natura 2000 Programme.[52]

The Habitats Directive

By the time the Commission came to propose the Habitats Directive in 1988 the Member States had had almost 10 years' experience of the Birds Directive's implementation. For many of them, this had not been a particularly happy experience: they were belatedly waking up to the legal significance of the habitat protection parts of the Directive, as well as the hunting control measures, often via repeated trips to the Court of Justice in Luxembourg.[53] Some saw the Commission's enforcement of the Birds Directive as "pernickety" and many clearly saw the Habitats Directive negotiations as an opportunity for the Member States to retake control in various respects.[54] In any event, by the time the Habitats Directive came to be negotiated, the Member States were very wise to the potential implications of strict habitat protection measures at EU level. The negotiations regarding the Habitats Directive ultimately took more than three years,[55] spanning seven Council Presidencies in total (Greece, Spain, France, Ireland, Italy, Luxembourg and the Netherlands), albeit most of the important work was done during the French, Luxembourg and Dutch Presidencies.[56]

Greek presidency: July–December 1988

The text of the draft Habitats Directive was formally proposed to the Council on 16 August 1988,[57] during a Greek presidency of the Council. By this time, the EU (or EEC, as it was then) had grown to comprise 12 Member States, with Greece, Portugal and Spain having joined the nine who had earlier negotiated the Birds Directive. The remainder of 1988 was marked by various objections in principle to the proposal by certain Member States, including an attempt to kill off the proposal altogether on the basis (or perhaps better the *pretext*) that Member States should focus on better implementation of the Bern Convention rather than embarking on the negotiation of a new instrument of EU law.[58]

The Commission's Directive proposal went beyond the establishment of protected areas by seeking to secure protection for important areas *outside* the Natura 2000 network in various ways, as summarised in Table 11.3.

Focusing on the operational provisions of the proposal: under Article 8 of the Commission's original proposal, outside protected areas,

> Member States shall also strive to avoid pollution or deterioration of natural and semi-natural habitats....Member States shall furthermore take

ABLE 11.3 Excerpts from the text of the Commission's Habitats Directive proposal (August 1988)[59] versus comparable text from the adopted version of the Habitats Directive (May 1992)[60]

ext of Commission's Habitats Directive proposal 4ugust 1988)	*Text of adopted version of Habitats Directive (May 1992) following negotiations*
reamble	Preamble
Vhereas such classification of special protection areas should not mean that excessive exploitation of nature should be allowed in non-protected areas and whereas appropriate land management also outside the classified areas should ensure the sustainability of the biological systems connecting these areas;	Whereas land-use planning and development policies should encourage the management of features of the landscape which are of major importance for wild fauna and flora;
Vhereas there is a need to protect distinct features of the landscape;	
..]	
Vhereas it might be necessary to take initiatives to promote common or coordinated management of transfrontier areas of major importance for the conservation of threatened species of wild fauna and flora and the maintenance of wilderness areas, [...]	
rticle 8	Article 3
[Outside special protection areas] Member States shall also strive to avoid pollution or deterioration of natural and semi-natural habitats.	3 Where they consider it necessary, Member States shall endeavour to improve the ecological coherence of Natura 2000 by maintaining, and where appropriate developing, features of the landscape which are of major importance for wild fauna and flora, as referred to in Article 10.
Member States shall furthermore take measures to ensure the protection of distinct features of the landscape, including in particular those specified in accordance with Annex VII,[61] which are of outstanding local importance to wildlife.	Article 10
Member States may allow derogations, limited to the minimum area necessary, from paragraph 2 if their impact can be counterbalanced, so that they have no adverse effects on the sustainability of the biological systems connecting the special protection areas or on species whose conservation cannot be ensured solely through the designation of special protection areas.	Member States shall endeavour, where they consider it necessary, in their land-use planning and development policies and, in particular, with a view to improving the ecological coherence of the Natura 2000 network, to encourage the management of features of the landscape which are of major importance for wild fauna and flora.
	Such features are those which, by virtue of their linear and continuous structure (such as rivers with their banks or the traditional systems for marking field boundaries) or their function as stepping stones (such as ponds or small woods), are essential for the migration, dispersal and genetic exchange of wild species.

(Continued)

Text of Commission's Habitats Directive proposal (August 1988)	*Text of adopted version of Habitats Directive (May 1992) following negotiations*
Article 9 Member States shall assist the Commission in promoting: (a) common or coordinated management of transfrontier areas of major importance for the conservation of species of wild fauna and flora specified in accordance with Annexes I and II or for the protection of the threatened habitats specified in accordance with Annex IV; (b) the maintenance of wilderness areas throughout the Community; [...]	No equivalent

> measures to ensure the protection of distinct features of the landscape, including in particular those specified in accordance with Annex VII, which are of outstanding local important to wildlife.[62]

Here there was to be some flexibility, with the proposal providing for derogations from the obligation to take measures to protect distinct landscape features, any derogation to be limited to the minimum area necessary,

> if [the derogations'] impact can be counterbalanced, so that they have no adverse effects on the sustainability of the biological systems connecting the special protection areas or on species whose conservation cannot be ensured solely through the designation of special protection areas.[63]

During this early period in the Directive proposal's history, Stanley Johnson and Claus Stuffmann – two European Commission officials within what is now DG Environment, who played key roles in the history of the Habitats Directive – embarked jointly on bilateral negotiations with three Member States (the Netherlands, Spain and the UK) in advance of a crucial Environment Council meeting on 24 November 1988, which would determine whether the proposal would sink or swim.[64] The meeting with the Netherlands is of particular interest for present purposes.

The Commission's note of discussions with the Netherlands complicates any characterisation of the Habitats Directive as a "Dutch directive," a perception which appears to have developed in certain quarters over time.[65] While the Netherlands undoubtedly played an important role in negotiating the Directive, the Directive's central elements (Natura 2000 network and parallel species protection regime) were already in place in the Commission's original proposal, and the Netherlands indeed complained that it had not been consulted regarding the Commission's draft directive before the text was proposed to the Council.[66] Moreover, an important "network" element of the proposal – that is, the protection afforded to features such as corridors outside protected areas – was watered down during negotiations, such that the text finally adopted imposes weaker obligations in this respect than were contained in the Commission's original proposal, albeit the Netherlands repeatedly sought to strengthen this aspect of the Directive.

Thus, rather than the Netherlands having "uploaded" its concept of a European Ecological Network to the EU level,[67] a better characterisation of what happened is that the Netherlands tried, but failed, to upload its concept. As one of France's negotiators of the Habitats Directive, Henri Jaffeux, comments, "If there is a country that [was] not satisfied out of the [Habitats Directive] negotiation, it is indeed the Netherlands,"[68] because the Directive as finally adopted was less ambitious than the Netherlands had sought, in particular in respect of defining and protecting corridors between sites.

These failed ambitions, and the fact that in 1993 the Fifth Environmental Action Programme contained only a relatively weak endorsement regarding corridors

> An interrelated network of habitats, based on the concept for Natura 2000, should be created through the restoration and maintenance of habitats themselves and of corridors between them. The creation and maintenance of this network will be very much dependent on how carefully transport, agricultural and tourist policies are shaped and pursued in the future[69]

led the Netherlands, shortly after the Habitats Directive was adopted, to continue pushing its European Ecological Network (EECONET) concept via a major conference in Maastricht in November 1993.[70] As Jaffeux explains, the aim of the Maastricht conference "was to overcome the Habitats Directive and the EU by proposing an ecological network for the whole of Europe based on a regional implementation of the Rio Convention on Biological Diversity as a pan-European strategy of biodiversity...."[71] This ultimately led, in 1995 in Sofia, to the endorsement by 54 countries of the Pan-European Biological and Landscape Diversity Strategy, under the auspices of the Council of Europe and others.[72]

Returning to the Habitats Directive negotiations, while an attempt was made to derail the directive proposal at the Environment Council meeting on 24 November 1988, with both the UK and Spain contesting the need for EU action in the area, this did not succeed.[73] To give a sense of the times, the UK was represented at this meeting by a junior minister because its staunchly Eurosceptic Secretary of State for the Environment, Nicholas Ridley, refused to attend EU Council meetings; Ridley had previously declared that there would be a Habitats Directive "over my dead body"![74] Nonetheless, the directive proposal proceeded. Shortly thereafter, in December 1988, the UK and Denmark tried to undermine the need for a Habitats Directive by pushing for the habitat protection provisions of the Bern Convention to be refined, and arguing that this should be the focus and that the EU should not duplicate effort.[75] While this attempt ultimately led to the adoption of a Resolution and three Recommendations by the Standing Committee of the Bern Convention in June 1989, it did not succeed in halting progress towards a Habitats Directive.[76] Interestingly, amongst the Recommendations agreed in June 1989 was one that advised countries to take steps to designate areas of special conservation interest, which ultimately led to the Bern Convention's Emerald Network.[77] Thus, as the present author has noted elsewhere, what became the Emerald Network was initially born of an attempt to prevent the development of the Habitats Directive.[78]

Spanish presidency (January–June 1989)

The new year brought a new Commissioner for the Environment, the Italian Carlo Ripa di Meana,[79] as well as a new presidency of the Council, held by Spain for the first half of 1989. While Ripa di Meana had not wanted the environment portfolio, having been happy in his job as EU Commissioner for Culture,[80] on appointment as Environment Commissioner he "at once made it plain that he intended to use the Commission's legal powers and his own considerable political skills to introduce a radical programme of environmental improvement in the Community."[81] Surprisingly, Spain's reticence regarding the Habitats Directive proposal appeared to have receded in a short space of time, when it was revealed that the proposal would be one of Spain's environmental priorities during its six-month presidency of the Council.[82] Nevertheless, little progress was made regarding the Habitats Directive brief during the Spanish presidency, and it was not until the subsequent presidency of the Council, under France from July to December 1989, that progress really began to be made.

However, during the Spanish presidency countries began to focus less on issues on principle (i.e. whether there should be a directive *at all*) and instead began to engage with the substantive provisions of the proposal. Regarding the parts of the proposal relating to the protection of areas outside Natura 2000: at a Working Party on the Environment meeting on 3 February 1989,[83] Germany and Denmark expressed (unspecified) doubts about the need for a provision such as Article 8 (see Table 11.3). While this is speculation on the present author's part, in respect of Article 8(1) ("1. [Outside special protection areas] Member States shall also strive to avoid pollution or deterioration of natural and semi-natural habitats"), certain Member States might have tried to argue that such a provision would be duplicative, given the similar language in the second sentence of Article 4(4) of the Birds Directive ("Outside these protection areas, Member States shall also strive to avoid pollution or deterioration of habitats").

Several delegations, in particular Belgium, Spain, France, Greece, Italy and Portugal, reportedly thought that the aim in proposed Article 8(2) of protecting distinct landscape features was desirable, but expressed (again unspecified) doubts about the advisability of including such a provision in the directive.[84] Indeed, despite the "desirability" of such an aim, Belgium, Spain, France and Greece reportedly favoured deleting the reference to protecting distinct landscape features altogether, while Italy and Portugal wanted the language transferred to the recitals. The Commission's unofficial note of the meeting[85] records a slightly different story. According to the Commission, France, Germany, Italy, Greece and Ireland *really* wanted Article 8 to be deleted.[86] The UK, the Commission reported, felt Article 8(1) was too general but was satisfied to retain Article 8(2).[87] Contradicting the official account, the Commission reported internally that Denmark was content to retain Article 8(2); while Germany, for its part, reportedly feared overloading the directive.[88] The Commission emphasised the importance of corridors between protected areas but said it was ready, if necessary, to move Article 8 to the recitals.[89]

In addition to these Article 8 provisions, Article 9 of the proposal aimed to oblige Member States to assist the Commission in promoting, amongst other things: (a) common or co-ordinated management of transfrontier areas of major importance for the conservation of species in Annexes I and II and for the protection of habitats in Annex IV; and (b) the maintenance of wilderness areas throughout the Community.[90] At the Working Party on the Environment's meeting on 3 February 1989, Germany, France, Greece and Ireland reportedly expressed doubts about including such provisions within the directive.[91] In its unofficial note of the meeting, the Commission records a slightly different story: no objection to common/coordinated management of transfrontier areas, but a general reservation regarding the idea of the maintenance of wilderness areas.[92]

French presidency (July–December 1989)

While only minor progress was made on the Habitats Directive proposal during the Spanish presidency, the opposite was true of the French presidency under Environment Minister Brice Lalonde (ex-head of Friends of the Earth France), which held office for six months from July 1989.[93] In addition to French leadership, several political developments around this time also helped to put the wind in the sails of the Habitats Directive proposal. First was the rise of the Greens. As the *New York Times* reported in March 1989, "Across Western Europe, mainstream politicians are scrambling to embrace the cause of the endangered environment as small ecological parties register startling electoral triumphs," causing Margaret Thatcher and French President François Mitterand to perform "sharp U-turns...to put environmental concerns at the top of their policy agendas."[94] Similarly, support for the environment was on the rise in Germany, with the Greens growing in support to comprise, by April 1989, the country's third political force, thus holding the balance between the two big parties.[95] Second was the departure of Nicholas "over my dead body" Ridley in the UK: when a cabinet reshuffle took place in July 1989, Ridley was moved sideways by Thatcher into the Department of Trade and Industry, to be replaced as Environment Secretary by the pro-European Chris Patten,[96] thus giving the Department of the

Environment the European perspective that the House of Lords had called for shortly before.[97] Third was Denmark's change of heart: as Veit Koester, Denmark's senior negotiator of the Habitats Directive, explained:

> at a certain time during the negotiations...the head of department, Ministry of Environment, he said to me, "We have to give in, we have to change our position and we have to go to be positive to the negotiations, we cannot stick to our idea that the EEC shall not interfere with our legislation in that field."[98]

Claus Goldberg, who worked under Koester and acted as Denmark's main negotiator once substantive negotiations over the text began, reflects:

> I do think that the Danish position was that we were more hesitant in the beginning, and when we found out, no, we can't escape, we can't use existing Danish legislation, we need to go in through this directive, we were pretty positive.[99]

The incoming French presidency, led by its senior negotiator Alain Megret, was keen to revise the draft directive and sought to embed a totally different approach to the adversarial one Member States had complained of regarding the Birds Directive's implementation.[100] Having previously worked for ten years on the free movement of medical specialists, dentists and pharmacists, following mutual recognition of their qualifications, Megret had a positive experience of a collaborative, consensual approach to EU negotiations, and this was the type of approach he felt was essential to imprint on the Habitats Directive negotiations.[101] Marcel Jouve, then head of the International and European Affairs division of the Nature Protection Directorate at France's Department of the Environment, undertook a tour of capitals at the start of the French presidency and comments that this "was probably very important in the negotiation process, as it was a strong signal of our determination to make progress, and of our openness to discuss our and [others'] ideas and find compromise formulas."[102] The French presidency built on these bilateral consultations with Member States to produce an overhauled draft of the Directive, having assembled a broad consensus around key points.[103]

As recorded above, many Member States had expressed doubts about including any provision in the Habitats Directive relating to the protection of landscape features outside the Natura 2000 network. Thus, it could have come as little surprise to the Commission when the French presidency suggested an approach that would significantly weaken the obligation on Member States in respect of areas outside Natura 2000 sites and proposed the deletion of Annex VII, which listed the distinct landscape features requiring protection, albeit retaining an amended version of the underlying obligation, as a new Article 10, second paragraph (see Table 11.4).[104] The Commission adopted the French approach at the meeting of

TABLE 11.4 Excerpts from the text of the French presidency's Habitats Directive compromise proposal (October 1989)[108] versus comparable text from the adopted version of the Habitats Directive (May 1992)[109]

Text of French presidency's Habitats Directive compromise proposal (October 1989)	*Text of adopted version of Habitats Directive (May 1992) following negotiations*
[No preamble]	Preamble Whereas land-use planning and development policies should encourage the management of features of the landscape which are of major importance for wild fauna and flora;
Article 10 Member States shall also strive to avoid pollution or deterioration of natural and semi-natural habitats outside the classified special protection areas referred to in Article 7. Member States shall furthermore strive to ensure the protection of natural features which are of outstanding local importance to wildlife and which allow species to live normally in their areas of distribution such as hedges, woods, thickets, ponds, etc	Article 3 3. Where they consider it necessary, Member States shall endeavour to improve the ecological coherence of Natura 2000 by maintaining, and where appropriate developing, features of the landscape which are of major importance for wild fauna and flora, as referred to in Article 10. Article 10 Member States shall endeavour, where they consider it necessary, in their land-use planning and development policies and, in particular, with a view to improving the ecological coherence of the Natura 2000 network, to encourage the management of features of the landscape which are of major importance for wild fauna and flora. Such features are those which, by virtue of their linear and continuous structure (such as rivers with their banks or the traditional systems for marking field boundaries) or their function as stepping stones (such as ponds or small woods), are essential for the migration, dispersal and genetic exchange of wild species.

the Working Party of the Environment on 28 and 29 September 1989, the meeting notes recording that:

> In order to take account of the concern voiced by several delegations....[the Commission] suggested a more flexible wording, specifying that Member States should encourage the implementation of the measures needed to

> ensure protection of the features of the rural area, insofar as these had a direct effect on the protection of habitats.[105]

On the same day that the Working Party was thus considering weakening the proposed level of protection applicable to important areas outside the Natura 2000 network, the Dutch MEP Hemmo Muntingh presented the European Parliament's first report on the directive proposal, welcoming the draft directive's approach to such areas:

> ...one of the major positive features of the Habitats Directive is its attempt to prevent and eliminate the fragmentation of nature....It is...imperative to maintain or develop habitats of a sufficient size and to maintain and encourage migration routes and natural links between main propagation areas and smaller habitats. In short, the aim must be to establish a continuous ecological infrastructure.[106]

Despite having effectively conceded weaker wording on the subject, the Commission maintained its belief in the importance of this aspect of its proposal, noting in October 1989:

> The Commission is fully convinced that the Natura 2000 Network will only form a viable ecological infrastructure if the special protection areas are duly connected by linking corridors constituted by those natural features in the wider countryside that are of paramount importance for the maintenance or recovery of the species covered by the directive in their natural areas of distribution. These features will typically comprise hedges, ponds, streams, groves etc. It will not be acceptable for the Commission to limit the Community's effort to save its natural heritage to measures geared to a number – however impressive – of isolated islets.[107]

However, when the French presidency presented its compromise proposal on 13 October 1989,[110] the aspects of the directive relating to areas outside Natura 2000 had indeed been weakened. Article 9 of the Commission's original proposal, which related to common or coordinated management of transfrontier areas of major importance for the conservation of species or for the protection of threatened habitats, and the maintenance of wilderness areas throughout the Community, had been deleted altogether. The requirement to take measures to ensure the protection of distinct features of the landscape, including in particular those specified in accordance with Annex VII (Article 8(2) of the Commission's original proposal), was gone too. While the French presidency's proposal retained (as Annex VI) an annex entitled "Protection of natural features enabling species to live normally in their area of distribution," the underlying obligation was not an obligation of protection: "Article 19. Particular attention shall be paid to research and work on the subjects specified in accordance with Annex VI, and transboundary co-operative research shall be encouraged."

What remained in the French presidency's proposal was brief (the first paragraph of Article 10 below effectively retains the language of Article 8(1) of the Commission's original proposal (see Table 11.3), which itself echoes the language of the second sentence of Article 4(4) of the Birds Directive, as noted above):

> Article 10
>
> Member States shall also strive to avoid pollution or deterioration of natural and semi-natural habitats outside the classified special protection areas referred to in Article 7.
>
> Member States shall furthermore strive to ensure the protection of natural features which are of outstanding local importance to wildlife and which allow species to live normally in their areas of distribution such as hedges, woods, thickets, ponds, etc

Certain Member States pushed at this time to strengthen this aspect of the directive. The Netherlands suggested in November 1989, for example, that "in the interests of the [Natura 2000] network's coherence, certain features connecting the various areas, such as rivers, hedges, etc., should also be included" within the network.[111] When this suggestion was not taken up, the Netherlands and Germany suggested that the above text of Article 10 should be supplemented as follows: "In applying the provisions of this Article, Member States shall give particular attention to the areas surrounding protected areas and to the natural features that serve as links between protected areas."[112] Again, this suggestion was not adopted.[113]

Despite its by this time generally favourable disposition towards the Habitats Directive proposal, in November 1989, the UK had the following to say regarding connectivity:

> As regards the question of providing a degree of protection outside the network of protected sites, many member states have expressed their opposition to the inclusion of any measure of this kind in the directive, on the basis it would seem that it is enough to get the site designation procedures in place. It seems unlikely therefore that anything beyond an obligation expressed in fairly general terms will feature in the final text of the directive. I recognise that this is less than the [House of Lords' select] committee may have sought, but perhaps the important thing is that the subject is appearing in the draft directive at all.[114]

On the contrary: experience with legal instruments would suggest that a weak or "woolly" provision that does not impose an enforceable obligation is little better than no provision at all, and the Member States would undoubtedly have known this.

Irish presidency (January–June 1990)

During the Irish presidency, the Commission was still apparently holding out hope for protecting areas outside Natura 2000 via the Directive: responding in January 1990 to an NGO's points regarding the Directive, a senior DG XI official wrote: "The Commission is attaching great importance to this aspect and will stand very firm against any attempt to limit the scope of the Directive to the mere network of special protection areas."[115]

Thus, when the Commission was required by the European Parliament to insert certain missing annexes formally into its proposal, the Commission took the opportunity to insert as Annex VII a list of "types of distinct features of outstanding local importance to wildlife" which the Member States would need to strive to protect, including, for example, valley corridors, coastal staging posts, bodies of water, fallow land and wooded farmland (including hedgerows).[116] Belgium, Spain, France and Luxembourg were in favour of deleting this annex "in view of the difficulty involved in applying it"; in contrast, Germany and the Netherlands were in favour of retaining it, but with more detail.[117] NGOs similarly picked up on a perceived lack of detail, with the Council for the Protection of Rural England writing to the Commission that

> We are concerned that the vague categories in the latest draft of Annexe VII...are so imprecise as to make their protection impossible...We fear that this may be a prelude to the removal or emasculation of Article 8 of the Directive [on protecting areas outside Natura 2000, from the Commission's original proposal].[118]

Italian presidency (July–December 1990)

During the Italian presidency, the Netherlands remained particularly committed to the idea of protecting areas outside Natura 2000, noting in a letter to the Commission of 16 October 1990 that

> An ecological frame of reference may consist of a European ecological network, composed of core areas of European importance and, in my opinion, of ecological corridors, enabling species to migrate from one core area to another...If desired, the Netherlands is willing to make available its know-how in realizing ecological networks...It is not impossible that in some cases we have to conclude that entirely new nature areas are to be developed in order to realise ecological coherence. In my opinion it is therefore necessary that the Habitats Directive pays attention to nature development in addition to conservation and restoration.[119]

Here we see an *attempt* by the Netherlands to "upload" the concept of its own National Ecological Network to EU level,[120] but this attempt was not ultimately successful given the limited protection finally afforded to areas outside Natura 2000, as shown below.

Luxembourg presidency (January–June 1991)

In the Luxembourg presidency's compromise proposal of 28 February 1991, the draft retained a fairly generic obligation – Article 8(1) – regarding the protection of important areas outside Natura 2000:

> Generally Member States shall strive to avoid pollution or deterioration of natural habitats and habitats of species located outside the special areas of conservation, in so far as they would have a significant effect having regard to the objectives of the present directive.[123]

However, the draft did not stop there, with a new provision having been added, providing:

> Article 8(2) To that effect, Member States shall in particular take into account, in the framework of their land-use planning and development policies, the features of the landscape of outstanding importance for the conservation of wild fauna and flora, such as those mentioned in annex IV, which allow species to move normally within their natural range and which ensure the coherence of Natura 2000. They shall endeavour to promote the conservation of such features through, among other measures, management contracts.[124]

Moreover, the draft also contained another new provision on the subject, as Article 3(2): "Member States shall enhance the ecological coherence of Natura 2000 by maintaining and, where appropriate, developing ecological corridors between natural habitats,"[125] a provision likely inspired by the Dutch government's suggestion of October 1990.[126] Thus, protection for areas outside the Natura 2000 network had made something of a late comeback in the Luxembourg presidency's draft of February 1991.[127]

In March 1991, the UK's Environment Secretary Michael Heseltine[128] voiced support for some form of protection of the broader countryside via the Habitats Directive, saying that

> the [Habitats] Directive must safeguard the general character of the landscape. This will help provide links to prevent the special sites becoming isolated and therefore vulnerable. Nature conservation cannot succeed as a series of individual oases in land which leaves no other footholds for our native wildlife. In essence this requires a continuation of the good stewardship which farmers in many parts of Europe have exercised for generations.[129]

However, when it came later that month to discussing the detail of the protective measures proposed for the wider countryside, the UK, Belgium, Spain, France and Ireland expressed reservations. The Article 3(2) obligation to maintain and, where appropriate, develop ecological corridors was a particular problem: Spain argued that it was "impossible to impose legal obligations outwith the designated areas and that only a political compromise might be possible"; the UK argued that Article 8(2) covered the issue (see Table 11.5); and, true to form, Denmark, Germany and the Netherlands "considered this [Article 3(2)] concept essential for the coherence of the network" and argued that a definition of ecological corridor should also be included.[130] Similarly, in an internal paper

TABLE 11.5 Excerpts from the text of the Luxembourg presidency's compromise proposal (February 1991)[121] versus comparable text from the adopted version of the Habitats Directive (May 1992)[122]

Text of Luxembourg presidency's Habitats Directive compromise proposal (February 1991)	*Text of adopted version of Habitats Directive (May 1992) following negotiations*
[No preamble]	Preamble Whereas land-use planning and development policies should encourage the management of features of the landscape which are of major importance for wild fauna and flora;
Article 3 2. Member States shall enhance the ecological coherence of Natura 2000 by maintaining and, where appropriate, developing ecological corridors between natural habitats. Article 8 1. Generally Member States shall strive to avoid pollution or deterioration of natural habitats and habitats of species located outside the special areas of conservation, in so far as they would have a significant effect having regard to the objectives of the present directive. 2. To that effect, Member States shall in particular take into account, in the framework of their land-use planning and development policies, the features of the landscape of outstanding importance for the conservation of wild fauna and flora, such as those mentioned in annex IV, which allow species to move normally within their natural range and which ensure the coherence of Natura 2000. They shall endeavour to promote the conservation of such features through, among other measures, management contracts.	Article 3 3. Where they consider it necessary, Member States shall endeavour to improve the ecological coherence of Natura 2000 by maintaining, and where appropriate developing, features of the landscape which are of major importance for wild fauna and flora, as referred to in Article 10. Article 10 Member States shall endeavour, where they consider it necessary, in their land-use planning and development policies and, in particular, with a view to improving the ecological coherence of the Natura 2000 network, to encourage the management of features of the landscape which are of major importance for wild fauna and flora. Such features are those which, by virtue of their linear and continuous structure (such as rivers with their banks or the traditional systems for marking field boundaries) or their function as stepping stones (such as ponds or small woods), are essential for the migration, dispersal and genetic exchange of wild species.

around this time, the Commission restated its attachment to the concept of ecological corridors in Article 8(2) and noted that the related obligation was not of the same nature (i.e. not as strict) as those that would apply in the context of designated areas.[131]

However, several Member States retained reservations about Article 8:[132] at a meeting of the Working Group on 18 April 1991, Spain and Portugal reportedly "felt the scope of the Article was so general that it was impossible to implement it and proposed replacing it with a policy statement." Portugal in any event wanted Article 8(1) amended to specify that only *deliberate* deterioration would be captured. For their part, Germany and Denmark argued that the implementation of Article 8 did not pose any problems in view of the latitude allowed to the Member States for transposing it.

This division in views was also evident when on 3 May 1991 the delegations discussed the compromise text from the Luxembourg presidency regarding Annex IV (formerly Annex VII), which set out a list of the landscape features outside the Natura 2000 network requiring protection under Article 8.[133] This list had grown considerably as a result of proposed amendments in MEP Hemmo Muntingh's second report for the European Parliament, some of which the Commission had taken on board in producing a formally modified directive proposal in February 1991.[134] Most delegations expressed reservations regarding Annex IV, some arguing that its application would be difficult and that, in imposing obligations solely in respect of the features listed, the Annex would lead to differences of application in view of the inevitable gaps in coverage.[135] Belgium, Spain, France and the UK thought that Article 8 could prove sufficient on its own, while Denmark, Germany and the Netherlands emphasised the importance of retaining the annex to ensure network consistency.[136] Denmark, France, Ireland, Portugal and Germany even went so far as to suggest additions to the list of landscape features of outstanding importance in Annex IV.[137]

This battle over the protection of the broader countryside continued when, also on 3 May 1991, the Luxembourg presidency presented a revised draft of the directive, containing the following (newly inserted) definition of "ecological corridors," linked to the obligation, now in Article 3(3), to maintain and, where appropriate, develop ecological corridors linking the special areas of conservation: "'ecological corridors' means terrestrial or aquatic formations of the countryside or coherent features which, in combination, are essential for the migration, geographical distribution and genetic exchange of species of wild fauna and flora."[138] Discussing this on 16 and 23 May 1991, Belgium, Spain, France, Greece, Ireland and the UK favoured deleting the new definition, and those countries, with Portugal replacing Greece,[139] favoured deleting the underlying obligation in Article 3(3); several of these countries considered the concept was too vague and too broad in scope, while Spain considered that provisions protecting areas outside Natura 2000 would raise legal problems, and again suggested a "simple political statement of intent."[140] In contrast, Denmark, Germany, Italy, the Netherlands and Portugal favoured retaining the definition and making the Annex IV list of features of interest more precise to facilitate implementation[141]; moreover, those countries, minus Portugal, expressly favoured retaining the underlying obligation in Article 3(3).[142]

The revised draft also contained an amended version of the other major provision in respect of ecological corridors, Article 8:

> Member States shall ensure that in their land-use planning and development policies and, in particular, in their relevant programmes to harmonize agricultural practices with the requirements of maintaining the countryside, encouragement is given to conservation of features of the landscape as listed in Annex IV which serve as ecological corridors.[143]

The UK, supported by Belgium, Ireland and Luxembourg, expressed reservations regarding this provision and suggested returning to an earlier draft.[144] Greece thought that the provision should be optional, while Spain and Portugal proposed deleting the Article altogether, perhaps to be replaced by a statement of intent.[145] Denmark, Germany, Italy, the Netherlands and the Commission were in favour of retaining the provision.[146]

The Netherlands continued working hard to retain the idea of ecological corridors in the draft Directive, producing a working document on the concept for all delegations on 6 June 1991, which concluded "ecological corridors and stepping stones are essential for the ecological coherence of...Natura 2000."[147] However, when the Luxembourg presidency presented revised text on the subject on 12 June 1991, the Netherlands' hopes were dashed: the proposal involved deleting the definition of "ecological corridors," deleting Article 3(3), and deleting Annex IV, which listed important features outside Natura 2000. In place of all of this, the sole provision on the subject would be the following weak obligation, as Article 8:

> Member States shall endeavour, where they consider it necessary, in their land-use planning and development policies and, in particular, with a view to improving the ecological coherence of the Natura 2000 network, to encourage the management of features of the landscape which are of major importance for wild fauna and flora.
>
> Such features are those which, by virtue of their linear and continuous structure (such as rivers with their banks or the traditional systems for marking field boundaries) or their function as stepping stones (such as ponds or small woods), are essential for the migration, dispersal and genetic exchange of wild species.[148]

A majority of delegations reacted positively to this proposal, and this text is identical to Article 10 of the Habitats Directive as adopted.[149]

Dutch presidency (July–December 1991)

During its own presidency, the Netherlands had one more go at strengthening the draft Directive in support of corridors, etc. Article 8 now – in July 1991 – remained unchanged from the position agreed at the end of Luxembourg's presidency (see immediately above).[150]

In addition to sticking with this version of Article 8, the Dutch presidency's compromise proposals stuck relatively closely to the Luxembourg presidency's proposals of the previous month, in that the definition of "ecological corridor" was to be deleted and the Annex IV list of landscape features was to be dropped.[151] However, rather than deleting the Article 3(3) obligation, as the Luxembourg presidency had proposed, the Netherlands proposed amending Article 3(3) to read:

> Where they consider it necessary, Member States shall endeavour to improve the ecological coherence of Natura 2000 by maintaining and, where appropriate, developing features of the landscape which are of major importance for wild fauna and flora, as referred to in Article 8(1).[152]

This might have been better than nothing, but it was weaker than the former Article 3(2), which had provided, "Member States shall enhance the ecological coherence of Natura 2000 by maintaining and, where appropriate, developing ecological corridors between natural habitats."[153]

So the difference between the Dutch presidency's position and the position reached at the end of the Luxembourg presidency was essentially the inclusion of Article 3(3), since the Luxembourg presidency had proposed, finally, retaining only Article 8.[154] The majority of delegations were generally in favour of the Dutch presidency's proposal. Denmark and Portugal considered that the examples of landscape features in brackets should be deleted from Article 8, while France felt that they should be amended.[155] Spain maintained its reservation regarding the Article, which it considered should be replaced by a declaration of intent.[156] However, at a Committee of Permanent Representatives meeting on 25 September 1991, Spain indicated that it was prepared to agree to the proposed wording for Article 8, but asked for Article 3(3) to be deleted.[157] Internally, the Commission admitted that it was prepared to accept this,[158] but Article 3(3) was finally retained following Spain's agreement to its retention alongside Article 8,[159] with the same text as noted immediately above.[160]

Habitats Directive agreed

Interestingly, less than a week before this weak outcome was codified by agreement of the Habitats Directive on 12 December 1991, the Standing Committee of the Bern Convention adopted a Recommendation setting out a much more detailed, ambitious policy on the conservation of natural areas outside protected areas, emphasising once more the complex interplay between the Council of Europe's and the EU's policies on nature conservation.[161] As Claus Stuffmann commented at a conference the year after the Habitats Directive's adoption,

> not all the Member States were ready to accept Community rules stricter than Article 3(3) …The Commission's original proposal was clearer and more prescriptive. It is nevertheless to be hoped that the firm belief shared by all today that nature conservation can not only be achieved by establishing protection areas but that it is vital to promote it in the whole territory of the Community, will in time lead to changes in Community legislation that could prove necessary.[162]

Some 30 years later, there have been no such changes. However, that may of course change in the not-too-distant future, given the EU Biodiversity Strategy for 2030 and the Commission's announcement in June 2022 of its proposal for an EU Nature Restoration Law, the EU's first major legislative foray in the field of nature conservation since the Habitats Directive.

The future

The EU Biodiversity Strategy for 2030 is clear that much more needs to be done at the level of EU law and policy, both in terms of implementing existing laws and policies and developing new ones: "The EU has legal frameworks, strategies and action plans to protect nature and restore habitats and species. But protection has been incomplete, restoration has been small-scale, and the implementation and enforcement of legislation has been insufficient."[163] The Commission continues:

> To put biodiversity on the path to recovery by 2030, we need to step up the protection and restoration of nature. This should be done by *improving and widening our network of protected areas* and by *developing an ambitious EU Nature Restoration Plan.*[164]

Both elements of this are relevant in the context of protecting areas outside the existing Natura 2000 network.

Improving and widening the network of protected areas

Regarding protected areas, the Strategy summarises its own "key commitments by 2030" as points 1–3 below. The present author's comments are noted below each point.

1 Legally protect a minimum of 30% of the EU's land area and 30% of the EU's sea area and integrate ecological corridors, as part of a true Trans-European Nature Network.[165]

The Strategy notes that "at least 30% of the land and 30% of the sea should be protected in the EU. This is a minimum of an extra 4% for land and 19% for sea areas as compared to today."[166] One might ask how this can be the case when Natura 2000 covers (only) 18% of the EU's land territory and 9% of its marine area. As the Commission explains,

> Today, 26% of the EU's land area is already protected, with 18% as part of Natura 2000 and 8% under national schemes. Of EU seas, 11% are protected, with 8% [now 9%] in Natura 2000 and 3% under additional national protection.[167]

Member States are to be responsible for designating the additional protected and strictly protected areas, which can either help to complete the Natura 2000 network or be under national protection schemes.[168]

Pursuant to the Strategy, the Commission, working with Member States and the European Environment Agency, was to put forward in 2020 criteria and guidance for identifying and designating additional areas, including a definition of strict protection, as well as for appropriate management planning, which criteria and guidance the Commission would aim to agree with Member States by the end of 2021. Ultimately, the guidance and criteria emerged as a Commission Staff Working Document in January 2022.[169]

It is important to note that the Staff Working Document is expressly stated to be non-binding and aims to help "Member States fulfil the *political commitment* expressed in the above-mentioned Council conclusions when they welcomed the protected areas targets of the EU Biodiversity Strategy" (emphasis added).[170] Thus, for now at least, the targets are a political rather than a legal commitment. That said, as the Strategy makes clear,

> Member States will [...] have until the end of 2023 to demonstrate significant progress in legally designating new protected areas and integrating ecological corridors. On this basis, the Commission will assess by 2024 whether the EU is on track to meet its 2030 targets or whether stronger actions, *including EU legislation*, are needed (emphasis added).[171]

Pursuant to the Commission's guidance,

> Each Member State is expected to submit to the Commission a list of existing protected areas (in addition to Natura 2000) which fulfil the criteria as well as an initial pledge for new areas to be designated explaining: (1)

> which criteria were used for their identification; (2) the scientific evidence that leads to their selection for designation; and (3) the mechanism that will be put in place to ensure their effective and adequate management and monitoring.[172]

In other words, sites put forward will not necessarily fall under the umbrella of specific EU legal protection and may remain under purely domestic protection, but something will need to be in place by way of adequate management and monitoring. Thus, nationally protected areas and other effective area-based conservation measures (OECMs) will only count towards the 30% target if they comply with the criteria set out in the Commission's guidance.[173]

In terms of corridors, the Strategy states that

> in order to have a truly coherent and resilient Trans-European Nature Network, it will be important to set up ecological corridors to prevent genetic isolation, allow for species migration, and maintain and enhance healthy ecosystems. In this context, investments in green and blue infrastructure and cooperation across borders among Member States should be promoted and supported, including through the European Territorial Cooperation.[174]

The 30% targets for protected areas on land and at sea are said by the Commission's guidance document to be "part of a more general objective of building a truly coherent Trans-European Nature Network, integrating ecological corridors."[175] In other words, the designation of corridors may[176] count towards the 30% targets. The Commission also states that, in reviewing the pledges for designations of *protected areas*, the Commission and the EEA will together "assess the overall coherence of the Trans-European Nature Network, proposing corrective action as necessary," adding that "the existence of ecological corridors and their functionality will be part of [this] assessment of coherence of the Trans-European Nature Network."[177]

2 Strictly protect at least a third of the EU's protected areas, including all remaining EU primary and old-growth forests.[178]

Given that the EU's expanded network of protected areas may comprise areas currently outside Natura 2000, including for example some corridors, this concept of strict protection may be relevant in the context of the present chapter. The Biodiversity Strategy for 2030 states that "at least one third of protected areas – representing 10% of EU land and 10% of EU sea [up from 3% and 1% currently] – should be strictly protected."[179] In terms of the ecosystems that are to benefit in particular from strict protection, the Strategy singles out "all the EU's remaining primary and old-growth forests," as well as "significant areas of other carbon-rich ecosystems, such as peatlands, grasslands, wetlands, mangroves and seagrass meadows [...], taking into account projected shifts in vegetation zones" and "areas of very high biodiversity value or potential."[180]

As envisaged by the Biodiversity Strategy, the Commission has provided a definition of strict protection in its guidance document, as follows:

> Strictly protected areas are fully and legally protected areas designated to conserve and/or restore the integrity of biodiversity-rich natural areas with their underlying ecological structure and supporting natural environmental processes. Natural processes are therefore left essentially undisturbed from human pressures and threats to the area's overall ecological structure and functioning, independently of whether those pressures and threats are located inside or outside the strictly protected area.[181]

The Commission also cross-refers to the IUCN's categories Ia (strict nature reserve), Ib (wilderness area), and II (national park) in this context, and notes that while categories Ia and Ib are largely in line with the objective specified in the Biodiversity Strategy, category II "allows for a process of zoning, in which strict protection does not necessarily apply to the whole protected area."[182]

Again, these targets relating to strict protection are clearly a *political* rather than a formal legal commitment, and the European Commission will similarly need to assess by 2024 whether stronger actions, including EU legislation, are needed.

3 Effectively manage all protected areas, defining clear conservation objectives and measures, and monitoring them appropriately.[183]

Here the Commission's guidance states that

> Without prejudice to existing obligations under the Birds and Habitats Directives, Member States are expected to put in place the necessary governance systems and allocate sufficient resources to ensure that all protected areas, in Natura 2000 or under national protection schemes, as well as all OECMs are adequately managed and monitored

and that "The setting up of conservation objectives and measures to deliver them is the first fundamental step to ensure the effectiveness of protected areas."[184] It suffices for present purposes to note that the process of defining conservation objectives and measures is ongoing for the existing Natura 2000 network, some 30 years after the Habitats Directive was adopted, such that significant resources and effort will be required to define the same for an expanded network in a timely fashion.

Developing an ambitious EU Nature Restoration Plan

This element of the Biodiversity Strategy is broad, encompassing a whole range of matters, amongst which is a commitment to put forward a proposal for legally binding EU nature restoration targets, which ultimately took the form of the proposal for an EU Nature Restoration Law.[185] While a full assessment of the proposed EU Nature Restoration Law is beyond the scope of this chapter, a brief review is necessary given that much of the proposed law goes to the question of areas outside the Natura 2000 network.

The first thing to note is that the proposed EU Nature Restoration Law is in the form of a Regulation – unlike the Habitats and Birds *Directives* – meaning that once adopted it will apply directly in the Member States' legal systems, requiring no national transposition measures. This removes one potential source of delay in implementation and should provide for greater consistency when it comes to application too.

The proposed Regulation requires Member States, by way of an overarching objective, to "put in place, without delay, effective and area-based restoration measures which together shall cover, by 2030, at least 20% of the Union's land and sea areas and, by 2050, all ecosystems in need of restoration."[186] Article 4 of the Regulation requires the restoration of certain terrestrial, coastal and freshwater ecosystems: namely, habitats listed in Annex I of the Regulation (all of which are habitat types listed in Annex I of the Habitats Directive), as well as the habitats of species listed in Annexes II, IV and V to the Habitats Directives, and the terrestrial, coastal and freshwater habitats of wild birds covered by the Birds Directive.[187] The measures to be put in place in respect of the habitats of species listed in Annexes II, IV and V to the Habitats Directives, and the habitats of species covered by the Birds Directive, include restoration measures to *enhance connectivity* (Article 4(3)). Further, restoration measures in respect of the habitats listed in Annex I to the Regulation must consider the need for *improved connectivity* between those habitat types (Article 4(5)). The restoration obligations under Article 4 apply to the specified habitat types inside *and outside* the Natura 2000 network, as the preamble underlines: "Member States should put in place restoration measures to ensure the recovery of protected habitats and species, including wild birds, across Union areas, *also in areas that fall outside Natura 2000*." As the Commission notes in respect of Annex I habitats under the Habitats Directive, for example: "The [Natura 2000] network covers approximately 34% of the surface of all Annex I habitat types, which means that about two thirds lies outside."[188]

Article 4(6) of the proposed Regulation requires the Member States to ensure a continuous improvement in the condition or quality of habitats subject to restoration and to prevent deterioration where good condition or sufficient quality has been reached. Article 4(7) provides as a general matter that "Member States shall ensure that areas where the habitat types listed in Annex I occur do not deteriorate." Again, the obligations in Article 4(6) and (7) apply to the specified habitat types inside *and outside* Natura 2000 sites. That this is so is clear from Article 4(8), which says

> *Outside Natura 2000 sites*, the non-fulfilment of [Article 4(6) and 4(7)] is justified if it is caused by: (a) force majeure; (b) unavoidable habitat transformations which are directly caused by climate change; or (c) a project of

overriding public interest for which no less damaging alternative solutions are available, to be determined on a case by case basis.

Overall, Member States must ensure that there is (a) an increase of habitat area in good condition for habitat types listed in Annex I of the Regulation until at least 90% is in good condition and until the favourable reference area for each habitat type in each biogeographic region of their territory is reached; and (b) an increasing trend towards the sufficient quality and quantity of the terrestrial, coastal and freshwater habitats of the species referred to in Annexes II, IV and V to the Habitats Directive and of the species covered by the Birds Directive. As noted above, these Article 4 rules apply to terrestrial, coastal and freshwater ecosystems, while similar rules are in place for marine ecosystems via Article 5. It is worth noting in this regard that the marine habitat types listed in Annex II to the Regulation, to which the restoration obligations *inter alia* apply,[189] include not only habitat types listed in Annex I to the Habitats Directive but also additional marine habitat types.

The EU Nature Restoration Law indeed goes much further, requiring restoration of urban ecosystems (Article 6); restoration of the natural connectivity of rivers and natural functions of the related floodplains (Article 7); restoration of pollinator populations (Article 8); and restoration of agricultural (Article 9) and forest (Article 10) ecosystems *in addition to* the habitats listed in Annex I of the Regulation, the habitats of species listed in Annexes II, IV and V to the Habitats Directives, and the habitats of species covered by the Birds Directive. Needless to say, this is all very ambitious, long overdue, and welcome, even if there, of course, remains some room for improvement.[190]

The proposal has now entered the next stage of the EU legislative process, in which the Council and the European Parliament are to negotiate the instrument's terms. But these negotiations cannot be too protracted, given the 2030 deadlines set by the Biodiversity Strategy and proposed in the instrument itself. Inevitably there will be resistance from some quarters. It is worth recalling, however, the huge inscriptions against the Birds Directive in the south of France at the time of that instrument's negotiation,[191] and the "awesome challenge to the sensitivities of member nations" reportedly presented by the Habitats Directive proposal.[192] Yet the Birds Directive was agreed "against all odds,"[193] and the feat was repeated a decade later with the Habitats Directive. The agreement and implementation of an ambitious EU Nature Conservation Law, going beyond the terms of the Habitats and Birds Directives to create strong, clear, time-bound obligations in respect of areas outside (as well as inside) Natura 2000, would be an important step towards the transformative changes that are needed to tackle the biodiversity and climate crisis. This will require imagination, courage, commitment, and a determined focus on the common good.[194]

Notes

1 European Commission, "Beyond 2010 – Options for EU biodiversity policy," MEMO/09/177, 23 April 2009.

2 Council Directive 92/43/EEC on the conservation of natural habitats and of wild fauna and flora, OJ L 206, 22.7.1992, p. 7, consolidated at http://data.europa.eu/eli/dir/1992/43/2013-07-01.
3 Council Directive 79/409/EEC of 2 April 1979 on the conservation of wild birds, OJ L 103, 25 April 1979, p. 1, since codified as Directive 2009/147/EC of the European Parliament and of the Council of 30 November 2009 on the conservation of wild birds, OJ L 20, 26 January 2010, p. 7, consolidated at http://data.europa.eu/eli/dir/2009/147/2019-06-26.
4 Based on data provided by the Member States the European Environment Agency regularly updates its Natura 2000 barometer https://www.eea.europa.eu/data-and-maps/dashboards/natura-2000-barometer.
5 For a review of the connectivity of protected areas in Europe and globally, see Santiago Saura et al, "Protected Area Connectivity: Shortfalls in Global Targets and Country-Level Priorities" (2018) 219 *Biological Conservation* 53–67.
6 For example, see the literature cited on pp. 37–38 and 84–85 of European Commission, "Commission Staff Working Document. Fitness Check of the EU Nature Legislation (Birds and Habitats Directives)," Brussels, 16.12.2016, SWD(2016) 472 final.
7 European Environment Agency, *The European Environment – State and Outlook 2020: Knowledge for Transition to a Sustainable Europe* (2019).
8 Caspar A. Hallmann et al, "More than 75 Percent Decline Over 27 Years in Total Flying Insect Biomass in Protected Areas" (2017) 12(10) *PLOS ONE* e0185809.
9 Laurianne Geffroy, "Where Have all the Farmland Birds Gone?" CNRS News, 21 March 2018, https://news.cnrs.fr/articles/where-have-all-the-farmland-birds-gone.
10 European Environment Agency, "State of Nature in the EU: Results from Reporting Under the Nature Directives 2013–2018," EEA Report No 10/2020 (2020), at Figure 3.1 and p. 72.
11 European Commission, "EU Biodiversity Strategy for 2030: Bringing Nature Back into our Lives," COM(2020) 380 final, Brussels, 20.5.2020.
12 European Commission, "Proposal for a Regulation of the European Parliament and of the Council on Nature Restoration," 22.6.2022, COM(2022) 304 final.
13 European Commission, "The State of Nature in the European Union," Report from the Commission to the Council and European Parliament. Brussels, 20.5.2015, COM(2015) 219 final, at p. 36.
14 Guy Pe'er et al, "How can the European Common Agricultural Policy Help Halt Biodiversity Loss? Recommendations by Over 300 Experts" (2022) *Conservation Letters* e12901.
15 European Court of Auditors, "Biodiversity on Farmland: CAP Contribution Has Not Halted the Decline," Special Report 13/2020 (2020).
16 Pe'er et al, supra n 14.
17 Case C-330/22 *Friends of the Irish Environment*, ECLI yet to be allocated.
18 European Commission, "Our Life Insurance, Our Natural Capital: An EU Biodiversity Strategy to 2020," 3.05.2011, COM(2011) 244 final (2011).
19 European Commission, "Communication from the Commission to the European Parliament, the Council, the European Economic and Social Committee and the Committee of the Regions. Green Infrastructure (GI) – Enhancing Europe's Natural Capital, COM/2013/0249 final (2013).
20 Council of the European Union, "Outcome of the 3441st Council of the EU meeting (Environment)," Brussels, 16 December 2015, 15380/15 (OR. en), PRESSE 84, PR CO 76.
21 For example, see Trinomics, "Supporting the Implementation of Green Infrastructure – Final Report," Report for DG Environment of the European Commission, May 2016.
22 Council of the European Union, "Council Conclusions on EU Action Plan for Nature, People and the Economy," Press release 387/17, 19 June 2017.

23 Supra n 11.
24 See https://wwfcee.org/naturaconnect.
25 Jonathan Verschuuren, "Connectivity: Is Natura 2000 Only an Ecological Network on Paper?," in: Charles-Hubert Born et al (eds.), *The Habitats Directive in its EU Environmental Law Context* (Routledge 2015), pp. 285–302.
26 Geert Van Hoorick, "Biodiversity Outside Protected Areas: An Outlaw Waiting to Be Saved?," in Charles-Hubert Born et al (eds.), *The Habitats Directive in its EU Environmental Law Context* (Routledge 2015), pp. 452–468.
27 Rachel Carson, *Silent Spring* (40th anniversary edition, Mariner Books 2002).
28 Jan-Henrik Meyer, "Saving Migrants: A Transnational Network Supporting Supranational Bird Protection Policy in the 1970s," in Wolfram Kaiser, Brigitte Leucht and Michael Gehler (eds.), *Transnational Networks in Regional Integration. Informal Governance in Europe 1945–83* (Palgrave Macmillan 2010), pp. 176–198.
29 Claus J. Stuffmann, "The Upcoming of the Birds Directive, 20 years with the EC Birds Directive" Proceedings from a conference on the Council's Directive on the Conservation of Wild Birds. The Danish National Forest and Nature Agency, Elsinore, Denmark (1999).
30 European Commission, "Proposal for a Council Directive on Bird Conservation," COM(76) 676 final, Brussels, 23 December 1976.
31 See Andrew L.R. Jackson, *Conserving Europe's Wildlife: Law and Policy of the Natura 2000 Network of Protected Areas* (Routledge 2018).
32 European Commission, supra n 30.
33 Stanley Cramp, "Bird Conservation in Europe: A Report," Prepared for the Environment and Consumer Protection Service, Commission of the European Communities (HMSO 1977).
34 European Commission, supra n 30, at paragraph I(5).
35 Jackson, supra n 31.
36 On which, see further Jackson, supra n 31.
37 See Geert Van Hoorick, supra n 26.
38 European Commission, supra n 30.
39 Supra n 3.
40 Article 2 provides:

> Member States shall take the requisite measures to maintain the population of the species referred to in Article 1 at a level which corresponds in particular to ecological, scientific and cultural requirements, while taking account of economic and recreational requirements, or to adapt the population of these species to that level.

41 Article 1 provides:

> 1. This Directive relates to the conservation of all species of naturally occurring birds in the wild state in the European territory of the Member States to which the Treaty applies. It covers the protection, management and control of these species and lays down rules for their exploitation. 2. It shall apply to birds, their eggs, nests and habitats.

42 All quotes in this paragraph from: Council of the European Communities, "Note: Proposal for a Council Directive on bird conservation [R/3265/76 (ENV 145)] – Outcome of the meeting of the Working Party on the Environment on Friday 11 March 1977," Brussels, 22 March 1977 (29.03). R/633/77 (ENV 46)(AGRI 174).
43 West Germany.
44 Article 2 of the original Birds Directive proposal read: "The Member States shall take the requisite measures to maintain the population of these species at a level compatible with ecological, economic, recreational and scientific requirements": see supra n 30.
45 Council of the European Communities, "Note: Proposal for a Council Directive on bird conservation [R/3265/76 (ENV 145)] – Outcome of the meeting of the

Working Party on the Environment on Thursday 28 and Friday 29 July 1977," Brussels, 31 August 1977 (05.09). R/1968/77 (ENV 113)(AGRI 550).

46 European Commission, supra n 30.

47 Supra n 3.

48 Council of the European Communities, "Note: Subject: Proposal for a Council Directive on bird conservation [R/3265/76 (ENV 145)] – Report from the Working Party on the Environment to the Permanent Representatives Committee," Brussels, 20 November 1977 (23.11). R/2741/77 (ENV 155)(AGRI 729).

49 The notes of prior meetings do not reveal the origins of this second sentence. At the meeting of the Working Party on the Environment held on 13 and 14 October 1977, Article 4(4) did not contain this second sentence, such that it must have been added to the proposal some time between then and 20 November 1977: compare ibid and Council of the European Communities, "Note: Proposal for a Council Directive on bird conservation [R/3265/76 (ENV 145)] – Results of the meeting of the Working Party on the Environment on Thursday 13 and Friday 14 October 1977," Brussels, 19 October 1977 (26.10). R/2441/77 (ENV 138)(AGRI 663).

50 Supra n 48.

51 Supra n 3. Note that this obligation has proven enforceable before the Court of Justice of the EU (CJEU): see the judgment in Case C-418/04 *Commission v Ireland*, EU:C:2007:780, in which Ireland was held to have failed to transpose and apply this obligation in practice.

52 Jan-Henrik Meyer, "Green Activism. The European Parliament's Environmental Committee Promoting a European Environmental Policy in the 1970s" (2011) 17 *Journal of European Integration History* 73–85, at 84.

53 See Jackson, supra n 31.

54 Ibid.

55 Political agreement was reached in December 1991, and the directive was formally adopted in May 1992: ibid.

56 Ibid.

57 See European Commission, letter dated 16 August 1988 from the Commission to the President of the Council, attaching Habitats Directive proposal COM(88) 381 final, Council document 8149/88, ENV 143, Brussels, 16 September 1988. The published version of the Commission proposal is dated 31 August 1988, but the proposal appears to have been submitted to the Council on 16 August.

58 See Jackson, supra n 31.

59 European Commission, supra n 57.

60 Supra n 2.

61 Annex VII was effectively blank in the Commission's proposal, stating only "The types of distinct landscape features to be specified under this Annex are those of outstanding local importance to wildlife." The text of the Annex was introduced to the negotiations later, and during the course of negotiations the numbering changed, such that it became Annex IV in later drafts, before ultimately being dropped before the directive was adopted.

62 European Commission, supra n 57, Article 8(1) and (2).

63 Ibid, Article 8(3).

64 See Jackson, supra n 31.

65 See Rüdiger Wurzel, "European Union Environmental Policy and Natura 2000," in Jozef Keulartz and Gilbert Leistra (eds.), *Legitimacy in European Nature Conservation Policy: Case Studies in Multilevel Governance* (Springer 2008), 267.

66 European Commission, internal DG XI letter dated 25 October 1988. Headed "Rapport de mission: Object: Mission à la Haye le 24/10/88. Discussion sur la proposition de directive habitats naturels (FFH)."

67 See Duncan Liefferink and Mariëlle van der Zouwen, "The Netherlands: The Advantages of Being 'Mr. Average,'" in Andrew Jordan and Duncan Liefferink (eds.),

Environmental Policy in Europe: The Europeanization of National Environmental Policy (Routledge 2004), 136–153.
68 Henri Jaffeux, Personal communication: E-mail dated 18 April 2013 from Henri Jaffeux to Andrew L.R. Jackson.
69 Fifth Environmental Action Programme, "Towards Sustainability: A European Community Programme of Policy and Action in Relation to the Environment and Sustainable Development," OJ C 138, 17 May 1993, p. 5.
70 Jaffeux, supra n 68.
71 Ibid.
72 Ibid.; and see Lawrence Jones-Walters, 'Pan-European Ecological Networks' (2007) 15 *Journal for Nature Conservation* 262–264.
73 See Jackson, supra n 31.
74 Ibid.
75 Ibid.
76 Ibid.
77 Standing Committee of the Bern Convention on the conservation of European wildlife and natural habitats, "Recommendation n° 16 (1989) of the standing committee on areas of special conservation interest," adopted by the Standing Committee on 9 June 1989.
78 Jackson, supra n 31.
79 European Commission, "Responsibilities of the members of the new Commission," Press release IP/88/825, 16 December 1988.
80 Jackson, supra n 31.
81 The Observer, "Profile: Playboy hero of the Greens – Carlo Ripa di Meana," 27 October 1991.
82 European Commission, DG XI (unit B.3.) file note dated 23 January 1989, headed "Object: Entretien du Commissaire Ripa di Meana avec le Président du Conseil Environnement le 24.1.89."
83 Council of the European Communities, "Outcome of the Working Party on the Environment on 3 February 1989," Brussels, 28 February 1989, 4708/89, ENV 25.
84 Ibid.
85 European Commission, note entitled "Aide-memoire, Réunion du groupe ENV/Conseil du 3/2/1989."
86 Ibid.; "souhaitent carrément."
87 Ibid.
88 Ibid.
89 Ibid.
90 Supra n 57.
91 Supra n 83.
92 European Commission, supra n 85.
93 Jackson, supra n 31.
94 James M. Markham, "Europe's politicians thinking green as concern grows on environment," The New York Times, 12 April 1989.
95 Henri de Bresson, "Differing shades of green," Guardian Weekly, 2 April 1989.
96 Michael White, "Major leaps up cabinet ladder: Howe is made deputy PM," The Guardian, 25 July 1989.
97 See Jackson, supra n 31.
98 Andrew L.R. Jackson, interview with Veit Koester, conducted on 22 August 2012.
99 Andrew L.R. Jackson, interview with Claus Goldberg, conducted on 31 August 2012.
100 Jackson, supra n 31.
101 Ibid.
102 Marcel Jouve, Personal communication: E-mail dated 8 April 2013 from Marcel Jouve, ex head of the International and European Affairs division of the Nature

Protection Directorate at France's Department of the Environment, to Andrew L.R. Jackson.

103 Jackson, supra n 31.

104 European Commission, DG XI (Unit B3) file note dated 19 September 1989, regarding a meeting between DG XI and French representatives to discuss the draft habitats directive, held on 13 September 1989 in Brussels.

105 Council of the European Communities, Outcome of proceedings of the Working Party on the Environment's meeting on 28 and 29 September 1989. Subject: Proposal for a Council Directive on the protection of natural and semi-natural habitats and of wild fauna and flora. Brussels, 9 October 1989. 8973/89, ENV 179, at p. 23.

106 European Parliament, Report, dated 29 September 1989, on the proposal for a Community directive on the protection of natural and semi-natural habitats and of wild fauna and flora (COM(88) 381 final – Doc. C 3–34/89). Drawn up on behalf of the Committee on the Environment, Public Health and Consumer Protection. Rapporteur: Mr. H.J. Muntingh. Series A, document A 3-0039/89, at p. 27.

107 European Commission, Document headed "III. Supplementary questions," attached to document headed "Speaking note" – for responding to Hemmo Muntingh's report for the European Parliament on the Habitats Directive proposal – with handwritten annotation at the top: "13.10.89 Parlement," at p. 3.

108 See infra n 110.

109 Supra n 2.

110 Council of the European Communities, "Cover note from General Secretariat to Delegations. Subject: Proposal for a Council Directive on the protection of natural and semi-natural habitats and of wild fauna and flora," attaching French presidency compromise text of Directive. Brussels, 13 October 1989, 9204/89, ENV 186.

111 Council of the European Communities, Outcome of the Working Party on the Environment on 9 November 1989. Brussels, 13 November 1989, 9862/89, ENV 206, p. 13.

112 Ibid., at p. 19.

113 Council of the European Communities, Background note from the Secretariat to the Committee of Permanent Representatives dated 13 November 1989, attaching revised draft directive. Brussels, 9847/89, ENV 205.

114 House of Lords, House of Lords' debate on 14 November 1989 regarding "Habitats and Species Protection: ECC Report," columns 1248–1279.

115 European Commission, Letter dated 4 January 1990 from DG XI to the Northamptonshire Wildlife Trust.

116 Council of the European Communities, Outcome of proceedings of Working Party on the Environment on 2 April 1990. Brussels, 24 April 1990, 5885/90, ENV 86. The obligation to protect was to arise by the addition of the words "and in particular those listed in Annex VII" to Article 10 of the French presidency's compromise proposal (the document suggests adding the language to Article 9, but that is clearly a typo and should read Article 10).

117 Council of the European Communities, ibid.

118 Council for the Protection of Rural England, Letter dated 6 April 1990 from the CPRE to DG XI's Nature Unit (B3). While the Commission's proposed Annex VII was made public on 30 March 1990 as a formal amendment of its original proposal [COM(90) 59 final], the French presidency's compromise draft of the Directive was of course not a public document, hence presumably why the NGO here fears the removal of Article 8 when this provision had in fact already been removed in the French presidency's draft.

119 Government of the Netherlands, Letter dated 16 October 1990 from the Nature Conservation, Environmental Protection and Wildlife Management Department of the Dutch government to DG XI.

120 Cf. Liefferink and van der Zouwen, supra n 67.

121 See infra n 123.

122 Supra n 2.
123 Council of the European Communities, Cover note from General Secretariat to Delegations dated 28 February 1991. Subject: Amended proposal for a Council Directive on the conservation of natural habitats and of wild fauna and flora – presidency compromise draft. Brussels, 4961/91, ENV 68.
124 Ibid.
125 Ibid.
126 Government of the Netherlands, supra n 119.
127 European Commission, DG XI document headed "Questions en suspens ("Defensive points")," behind tab "CONSEIL 18.3.1991" in DG Environment's archive.
128 Having lost out in the Conservative Party leadership election to John Major, Major then appointed Heseltine as Secretary of State for Environment in November 1990, replacing Chris Patten: Robin Oakley, "Party unity takes top priority in reshuffle" The Times, 29 November 1990.
129 Michael Heseltine, Speech made by Secretary of State Heseltine some time during the week beginning 11 March 1991, on the subject of wildlife protection.
130 Council of the European Communities, Outcome of proceedings of Working Party on the Environment on 21 March 1991. Brussels, 8 April 1991, 5572/91, ENV 107, at p. 13.
131 European Commission, "Questions en suspens 'Defensive points'": behind tab 18.3.1991 in DG Environment's archive.
132 Council of the European Communities, Outcome of proceedings of Working Party on the Environment on 18 April 1991. Brussels, 24 April 1991, 5845/91, ENV 122, at p. 3.
133 Council of the European Communities, Outcome of proceedings of Working Party on the Environment on 3 May 1991. Brussels, 16 May 1991, 6162/91, ENV 148.
134 European Commission, Modified proposal for a Council Directive on the conservation of natural and semi-natural habitats and of wild fauna and flora, COM(91) 27 final , 8 February 1991; and see European Parliament, Second report by the Committee on the Environment, Public Health and Consumer Protection on the proposal from the Commission to the Council for a directive on the protection of natural and semi-natural habitats and of wild fauna and flora (COM(88) 381 final + COM(90) 59 final – Doc. C 3–34/89). Rapporteur: Mr. H.J. Muntingh. A 3-0254/90.
135 Supra n 133.
136 Ibid.
137 Ibid.
138 Council of the European Communities, Cover note from General Secretariat to Delegations. Brussels, 3 May 1991, 5949/91, ENV 134.
139 Portugal's position appears quite convoluted here and may not be accurately recorded in the minutes.
140 Council of the European Commununties, Outcome of proceedings of Working Party on the Environment on 16 and 23 May 1991. Brussels, 28 May 1991, 6163/91, ENV 149, p.9.
141 Ibid.
142 Ibid.
143 Supra n 138.
144 Supra n 140.
145 Ibid.
146 Ibid.
147 Government of the Netherlands, Explanatory note from the Netherlands delegation concerning the concept of ecological corridors. Working document ENV/91/47, Brussels, 6 June 1991.
148 Council of the European Communities, Addendum to 6755/91 ENV 180, Brussels, 12 June 1991, 6755/91 ADD 1, ENV 180, at p.4. CEC, 1991a.

149 Ibid.

150 Council of the European Communities, Outcome of proceedings of the Working Party on the Environment on 10 July 1991. Brussels, 29 July 1991, 7700/91, ENV 237.

151 Ibid.

152 Ibid.

153 Council of the European Communities, Cover note from General Secretariat to Delegations dated 28 February 1991. Subject: Amended proposal for a Council Directive on the conservation of natural habitats and of wild fauna and flora – presidency compromise draft. Brussels, 4961/91, ENV 68.

154 Supra n 148.

155 Supra n 150.

156 Ibid.

157 Council of the European Communities, Outcome of proceedings of Permanent Representatives Committee on 25 September 1991. Brussels, 25 September 1991. 7992/2/91 REV 2, ENV 250.

158 European Commission, DG XI document headed "Points defensifs, Conseil Environnement, 1er octobre 1991."

159 European Commission, DG XI note dated 22 October 1991, headed "Rapport de mission," relating to the Environment Council meeting of 1 October 1991; and see the Habitats Directive, as adopted, supra n 2.

160 The only change was that "Article 8(1)" at the end of Article 3(3) had been amended to "Article 8": Council of the European Communities, Outcome of proceedings of Environment Council on 12 December 1991. Brussels, 20 December 1991, 10468/91, ENV 405.

161 Council of Europe, Recommendation n° 25 (1991) on the conservation of natural areas outside protected areas proper. Adopted by the Standing Committee of the Bern Convention on 6 December 1991; for more, see Jackson, supra n 31.

162 Claus Stuffmann, "The European Community and EECONET,' in G. Bennett (ed.), *Conserving Europe's Natural Heritage: Towards a European Ecological Network* (Graham & Trotman 1994), 203–210, at pp. 206–207.

163 Supra n 11.

164 Ibid. NB. There are other headings under the Biodiversity Strategy, including "Enabling transformative change" (comprising a new European biodiversity governance framework; stepping up implementation and enforcement of EU environmental legislation; and building on an integrated and whole-of-society approach) and "The European Union for an ambitious global biodiversity agenda."

165 Ibid.

166 Ibid.

167 Ibid.

168 Ibid.

169 European Commission, "Criteria and Guidance for Protected Area Designations" (Staff Working Document) 28.1.2022, SWD(2022) 23 final. In terms of its having been agreed with Member States, the document states that "This Commission Staff Working Document reflects the wide consensus reached as a result of the discussions held within the framework of the Nature Directives Expert Group (NADEG). It takes into account the work done in other groups, such as the Marine sub-Expert Group and the Working Group on Forests and Nature, on definitions (e.g. 'old growth forest') or specific criteria (e.g. for the marine environment). All these groups include representatives of Member States and stakeholders."

170 Ibid.

171 Supra n 11.

172 Supra n 169.

173 Ibid.

174 Supra n 11.
175 Supra n 169.
176 As the Commission's guidance states, "Parts of these ecological corridors will fulfil the criteria described above for protected areas and should therefore be counted towards the corresponding targets in the strategy. Many of them, however, are too small to be manageable as protected areas."
177 Supra n 169.
178 Supra n 11.
179 Ibid.
180 Ibid.
181 Supra n 169.
182 Supra n 169. That said, this does not necessarily seem to be a problem, given that elsewhere in the same guidance the Commission states "A protected area may be strictly protected in its entirety, but it is also possible that the area under strict protection is only a part of a wider protected area."
183 Supra n 11.
184 Supra n 169.
185 Supra n 12.
186 Ibid.
187 Ibid.
188 European Commission, Commission Staff Working document: Impact assessment accompanying the proposal for a Regulation of the European Parliament and of the Council on nature restoration, Part 1/12, Brussels, 22.6.2022, SWD(2022) 167 final.
189 Restoration obligations in Article 5 also apply to the marine habitats of species listed in Annex III to the Regulation and in Annexes II, IV and V to the Habitats Directive and to the marine habitats of wild birds covered under the Birds Directive.
190 For some suggestions as to how the proposal might be strengthened, see EEB, "EU nature restoration law: huge opportunity to fight biodiversity and climate crises – NGO reaction," 22 June 2022, https://eeb.org/eu-nature-restoration-law-huge-opportunity-to-fight-biodiversity-and-climate-crises-ngo-reaction/
191 See Jackson, supra n 31.
192 Marion Shoard, "Too precious, too Scottish to be left with the Scots" The Times, 30 June 1990.
193 Meyer, supra n 28, at p. 179.
194 In his speech to the UN Stockholm Conference in 1972, Sicco Mansholt asked delegates "not to be lured into repeating all the technical and scientific discussions but to draw the political conclusions from the situation" the world found itself in, which concerned the future liveability of the planet. "For this," he said, "we need imagination and courage." See Sicco Mansholt, 'Address by Mr. Sicco L. Mansholt, President of the Commission of the European Communities, to the Stockholm Conference', 8 June 1972, http://aei.pitt.edu/12980/1/12980.pdf

INDEX

Note: Page numbers followed by "n" denote endnotes.

For Product Safety Concerns and Information please contact our EU
representative GPSR@taylorandfrancis.com
Taylor & Francis Verlag GmbH, Kaufingerstraße 24, 80331 München, Germany

www.ingramcontent.com/pod-product-compliance
Ingram Content Group UK Ltd.
Pitfield, Milton Keynes, MK11 3LW, UK
UKHW020316270926
475702UK00050B/889
9780367275440